Book s

Advances in Industrial Control

Springer
London
Berlin
Heidelberg
New York
Barcelona
Hong Kong
Milan
Paris
Singapore
Tokyo

Other titles published in this Series:

Rashid M. Ansari and Moses O. Tadé

Nonlinear Model-based Process Control

Applications in Petroleum Refining

With 83 Figures

Springer

Rashid M. Ansari, PhD
Department of Chemical Engineering, Curtin University of Technology,
GPO Box U 1987, Perth 6845, Australia

Moses O. Tadé, PhD
Department of Chemical Engineering, Curtin University of Technology,
GPO Box U 1987, Perth 6845, Australia

ISSB 1430-9491

ISBN 1-85233-213-1 Springer-Verlag London Berlin Heidelberg

British Library Cataloguing in Publication Data
Ansari, Rshid M.
    Nonlinear model-based process control: applications in
    petroleum refining. - (Advances in industrial control)
    1.Petroleum - Refining 2.Nonlinear control theory
    I.Title II.Tade, Moses O.
    629.8'36
    ISBN 1852332131

Library of Congress Cataloging-in-Publication Data
Ansari, Rashid.
    Nonlinear model-based process control : applications in petroleum refining / Rashid M.
    Ansari and Moses O. Tadé.
        p. com. -- (Advances in industrial control)
    Includes bibliographical references.
    ISBN 1-85233-213-1 (alk. paper)
    1. Petroleum--Refining 2. Chemical process control. 3. Nonlinear control theory. I.
    Tadé, Moses O. II. Title. III. Series.
    TP690.3 .A57 2000
    665.5'3--dc21   99-047343

Typesetting: Camera ready by authors
Printed and bound by Athenæum Press Ltd., Gateshead, Tyne & Wear
69/3830-543210 Printed on acid-free paper SPIN 10731580

Advances in Industrial Control

Professor Dr -Ing M. Thoma
Institut für Regelungstechnik
Universität Hannover
Appelstr. 11
30167 Hannover
Germany

Professor H. Kimura
Department of Mathematical Engineering and Information Physics
Faculty of Engineering
The University of Tokyo
7-3-1 Hongo
Bunkyo Ku
Tokyo 113
Japan

Professor A.J. Laub
College of Engineering - Dean's Office
University of California
One Shields Avenue
Davis
California 95616-5294
United States of America

Professor J.B. Moore
Department of Systems Engineering
The Australian National University
Research School of Physical Sciences
GPO Box 4
Canberra
ACT 2601
Australia

Dr M.K. Masten
Texas Instruments
2309 Northcrest
Plano
TX 75075
United States of America

Professor Ton Backx
AspenTech Europe B.V.
De Waal 32
NL-5684 PH Best
The Netherlands

SERIES EDITORS' FOREWORD

The series *Advances in Industrial Control* aims to report and encourage technology transfer in control engineering. The rapid development of control technology has an impact on all areas of the control discipline. New theory, new controllers, actuators, sensors, new industrial processes, computer methods, new applications, new philosophies..., new challenges. Much of this development work resides in industrial reports, feasibility study papers and the reports of advanced collaborative projects. The series offers an opportunity for researchers to present an extended exposition of such new work in all aspects of industrial control for wider and rapid dissemination.

The last decade has seen considerable interest in reviving the fortunes of non-linear control. In contrast to the approaches of the 60s, 70s and 80s a very pragmatic agenda for non-linear control is being pursued using the model-based predictive control paradigm. This text by R. Ansari and M. Tadé gives an excellent synthesis of this new direction.

Two strengths emphasized by the text are:

(i)   four applications found in refinery processes are used to give the text a firm practical continuity;

(ii)   a non-linear model-based control architecture is used to give the method a coherent theoretical framework.

A key issue raised by this text concerns the ease with which realistic and accurate non-linear models can be generated for insertion into the non-linear model-based control architecture. The models for refinery processes have probably been reasonably well researched but many other areas may not be so fortunate. For example, non-linear bio-chemical reactors are one area in which it is difficult to devise phenomenologically based models, which have sufficient accuracy for control purposes.

We feel that this text is thoroughly topical and will be of considerable interest to the academic control community and also to the industrial engineer. The latter is likely to be very interested in the degree of enhanced performance that non-linear control can offer in real applications.

M.J. Grimble and M.A. Johnson
Industrial Control Centre
Glasgow, Scotland, UK

PREFACE

The work in this book entails developing non-linear model-based multivariable control algorithms and strategies and utilizing an integrated approach of the control strategy, incorporating a process model, an inferential model and multivariable control algorithm in one framework. This integrated approach was applied to various refinery processes, which exhibit strong non-linearities, process interactions and constraints and has been shown to produce good results for a range of refinery processes by improving the closed-loop quality control and maximizing the yield of high-value products. This makes non-linear model-based multivariable control an attractive alternative to linear methods. The generic model control (GMC) structure of Lee and Sullivan (1988) was selected for this research and practical work in this book which permits the direct use of non-linear steady-state and dynamic models and, therefore, provides the basic structure of the model-based controller.

The non-linear model-based control structure was further extended to permit the use of inferential models in non-linear multivariable control applications. A wide range of inferential models was developed, implemented in real-time and integrated with non-linear multivariable control applications. These inferential models demonstrate the improvement in the performance of closed-loop quality control and dynamic response of the system by reducing the long time delays.

In order to demonstrate the effectiveness of non-linear model-based control with regards to industrial application, a non-linear control strategy was developed for an industrial debutanizer and implemented in real-time. The non-linear control provided improved control performance of the product qualities. A steady-state model of debutanizer with approximate dynamics was used in the control strategies.

A complex multivariable control problem was solved by formulating the non-linear constrained optimization strategy for a crude distillation process. The heavy oil fractionator problem proposed by Shell Oil was selected for this work. The method uses the non-linear model-based controller, which considers the model uncertainty explicitly. The method is based on formulating the constrained non-linear optimization (NLP) programme that optimizes performance objectives subject to constraints. The model was built in MATLAB$^{®}$/SIMULINK$^{®}$ and the results were tested and compared with the results obtained from non-linear control techniques.

A constrained non-linear multivariable control and optimization strategy for handling the constraints was proposed and applied in real time to a semiregenerative catalytic reforming reactor section. An octane inferential model was developed and integrated with non-linear multivariable controller forming a closed-loop WAIT/octane quality control. A dynamic model of the catalytic reforming process was developed and used in this control application on the reactor section to provide target values for the reactor inlet temperature. It was shown that constrained non-linear multivariable control provided better disturbance rejection compared to traditional linear control. The optimization approach in this work provided a trade-off between the process outputs tracking their reference trajectories and constraint violation.

Finally a non-linear constrained optimization strategy was proposed and applied to a fluid catalytic cracking reactor-regenerator section using a simplified FCC process model. A dynamic parameter update algorithm was developed and used to reduce the effect of larger modelling errors by regularly updating the selected model parameters. The main advantage of the proposed non-linear controller development approach is that a single-time step control law resulted in a much smaller-dimensional non-linear programme as compared to other previous methods. Since the same model was used for optimization and control, it minimized the maintenance and process re-identification efforts, which was required for linear controller development as the operating conditions change.

One of the key objectives of the applied work in this book was to develop and implement in real-time the *non-linear model-based multivariable control* and *constraint optimization algorithms* on various refinery processes with strong non-linearities and process interaction. The implementation of these strategies and techniques in real-time has demonstrated an improved control performance over linear control, emphasizing the need of a high-performance model-based non-linear multivariable control system.

This book may serve as a concise reference for process control engineers interested in non-linear process control theory and applications. We tried to apply the non-linear control concepts and methods in real-time applications in petroleum refining industry and hope that the readers will find this book interesting and useful.

Rashid M. Ansari
Moses O. Tadé

ACKNOWLEDGEMENTS

It was always my desire and ambition to write a book in the field of advanced process control. The realization of this ambition demands a greater commitment and dedication to complete the work. During the last year, I realized that my ambition to write a book in control area turned into a great challenge for me and I have accepted that challenge. However, my ambition would have never been realized without the help of the following people:

I would like to express my sincerest gratitude to Dr. Moses O. Tadé for his invaluable guidance, helpful advice and consistent encouragement throughout the duration of this work. I am especially grateful to his thorough reading of several drafts of this book and his valuable corrections. My sincere thanks to Professor Terry Smith for encouraging me to continue working in the field of advanced control and optimization.

Many thanks to Dr. Weibiao Zhou and Dr. Peter L. Lee for their initial help in understanding the non-linear generic model control techniques. Thanks to my colleagues Ashraf A. Al-Ghazzawi and Dr. Talal Bakri for their help on setting up the MATLAB®/SIMULINK® control system.

I am also grateful to Majed A. Intabi, Manager, Riyadh Refinery who always admired my work and extended his support for any professional activity, which can contribute to the development of young engineers. Special thanks to Editors of Saudi Aramco Journal of Technology for publishing some of my work in the technical journal and Oliver Jackson, Editorial Assistant of Springer-Verlag for his help in completing the write-up of this book.

CONTENTS

LIST OF FIGURES

LIST OF TABLES

CHAPTER 1

INTRODUCTION

1.1 Non-linear Model-based Control

The need for high-performance control systems has accelerated over the past decade. Economic pressures, increased environmental and safety concerns, and a tighter integration of process units have all contributed towards this demand. At the same time, because of the increasing efficiency of process computers, the demand for robust and accurate control of non-linear industrial processes can now be realized by advanced control strategies. Among these strategies, non-linear model-based control (NMBC), a kind of multivariable control technique, has drawn considerable attention in recent years. Lee and Sullivan (1988) presented a framework for process controllers that allow for the inclusion of a non-linear mathematical model to approximate the plant behaviour. The so-called generic model control (GMC) formulation of the control law allows process models, linear and non-linear, to be embedded directly into the control structure, providing the potential for better process control due to more accurate process models being incorporated into the control structure.

Non-linear model-based control is an integrated approach for process analysis, control and optimization where the same steady state, non-linear, process model is used at each stage. The use of this type of model to predict the control action required to meet the control objectives can be expected to provide improved performance over simple, linear models as the model gives more precise information about the effects of manipulated variables (inputs) on the controlled variables (outputs). In addition, for multiple-input/multiple-output (MIMO) systems, the model accounts for interaction among the process variables.

Some non-linear model-based control algorithms also involve optimization, a pre-specified reference trajectory plays an important role in model-based control algorithm such as GMC. The control law is determined so that it drives the closed loop system to follow this reference trajectory. Therefore an optimization strategy for handling the constraints can be constructed in the model-based control structure which allows a balance to be sought between the process outputs tracking their reference trajectories and constraint violation. In this manner, constraint variables can be controlled to the constraint limits as well as to the rate of approach to the constraint limit. Slack variables defining the trajectory departure from the chosen specification curves can be added to the model-based control law for both the control variables and the constraint variables. The solution of the problem then becomes a non-linear constrained optimization, which minimizes a function of the slack variables. The choice of the weighting factors on the slack variables establishes the importance of the process constraints relative to the system setpoint tracking capability.

1.2 Motivation for this Book

The control strategies for chemical processes have been traditionally designed using simple, dynamic, linear, cause-and-effect models to describe the process. Although these models are sufficient for some processes, there are many other processes for which linear models do not provide an accurate basis for good control. The significance of non-linearities and the need for non-linear control has been recognized by both industrial practitioners and academic researchers. In addition, the presence of process non-linearities imposes limitations on the achievable control performance whenever linear controllers are used.

The first motivation for the work in this book in the direction of non-linear model-based multivariable control stems from the strong interest expressed by industries, in particular the process industries, to implement this technology as a means to improve operating efficiency and profitability. A second motivation, which drives many of the academic researchers, is to design a practical method and to implement that method in real-time to control the complex processes with strong non-linearities and process interaction. A third motivation is to bridge the widening gap between theoretical work and the industrial practice of advanced process control.

1.3 Objectives and Contributions

1.3.1 Objectives

The main objective of the research work in this book is to develop non-linear model-based multivariable control and optimization algorithms and strategies and to implement these strategies in real-time applications on various processes in the petroleum refinery. The emphasis is on an integrated approach to incorporate a process model, an inferential model and multivariable control algorithm in one framework. With regards to control, the objective is to provide improved control performance over linear model-predictive control, better process constraints handling and meeting the ever-changing process demands to bring the higher benefits to the refineries.

1.3.2 Contributions

1.3.2.1 Non-linear Control Theory and Development
This research work in non-linear model-based multivariable control is a direct contribution to process control theory and design. Previously non-linear model-based control strategies and algorithms were not developed and tested in real time for complex processes such as catalytic reforming or fluid catalytic cracking. An integrated approach which incorporates a process model, an inferential model and multivariable control algorithm in one control framework is also unique, providing improvement in closed loop quality control and dynamic performance of the system as compared to other linear techniques presently utilized. In addition, this integrated approach has several benefits: it eliminates the problems caused by linear methods of using different models for different functions and therefore reduces potential optimizer-controller mismatch; it also accounts for both non-linearities and interactions of a real process.

1.3.2.2 Practical Applications in Industry
This research and development work described in the book has a great significance with regards to its applications in industry. The benefits of applying the non-linear model-based multivariable control on refinery processes are very high and the payback on small projects can be less than six months. Also there are only a few applications of non-linear model-based control in industry and further development in this area will help industries to meet the demands of more challenging and competitive environment. In addition, steady state and dynamic process models developed and used in multivariable control applications can also be utilized for other purposes such as process trouble-shooting and for the assessment of product quality. With some modifications, these models can also be used in a simulator environment to train the operators on various processes.

1.4 Scope of the Book

The overall scope of this book entails developing the non-linear model-based multivariable control algorithms and constrained non-linear optimization strategies and utilizing an integrated approach of the control strategy, incorporating a process model, an inferential model and multivariable control algorithm in one framework. The GMC structure of Lee and Sullivan (1988) is selected for this research which permits the direct use of non-linear dynamic models and, therefore, provides the basic structure of the model-based controller. The scope of this study is summarized as follows:

1. To extend the non-linear model-based control structure to permit the use of inferential models in non-linear multivariable control applications.
2. To develop a wide range of inferential models, implement these models in real-time and integrate with non-linear multivariable control applications to form a closed-loop quality control.

3. To develop non-linear control strategies for industrial debutanizer using steady-state model with approximate dynamics and to implement the control strategies in real time.

4. To develop a complex non-linear multivariable control structure by formulating the constrained non-linear optimization method that optimizes the performance objectives for crude distillation process subject to constraints. Shell heavy oil fractionator was selected for this purpose, the process dynamic model was built by process identification tests and the controller considers the model uncertainty explicitly.

5. To develop a constrained non-linear optimization algorithm for the reactor section of semi-regenerative catalytic reforming process. To develop a dynamic model of the catalytic reforming process and an inferential model to predict the reformate octane. Integrate all these models in non-linear model-based control structure and implement control strategies in real-time.

6. To extend the application of constrained non-linear multivariable control and optimization to more complex process such as the fluid catalytic cracking (FCC) and to develop the dynamic parameter update algorithm for the FCC to reduce the effect of larger modelling errors by regularly updating the selected model parameters. To develop a simplified dynamic model of the process and to test the model in real-time to provide the initial targets for the non-linear controller. Finally, to apply the non-linear control in real-time and to compare these strategies with the DMC type controller built with the dynamic models obtained from process identification tests.

1.5 Book Overview

In this book, four important refinery processes were selected for the development of non-linear model-based multivariable control algorithm and its application in real-time. Figure 1.1 gives an overview of refinery processes considered in this book. The detailed process descriptions and non-linear control objectives and strategies are given in the chapters to follow.

The *crude oil fractionator* is a primary process unit of the refinery that performs the initial distillation of the crude oil into several boiling range fractions such as light ends, naphtha, kerosene, light gas oil (LGO), heavy gas oil (HGO) and reduced crude. Light ends and naphtha leave the fractionator overhead as vapours and liquid streams, respectively. Kerosene, LGO and HGO are removed as sidedraw products. Reduced crude leaves as the bottom product to vacuum unit for further processing.

The *debutanizer* column receives the overhead liquid stream, unstabilized naphtha, from the crude oil fractionator. The debutanizer fractionates the naphtha such that the light ends are removed from the top and debutanized naphtha is removed from the bottom and directed to catalytic reformer for further processing.

The *catalytic reformer* receives the stabilized naphtha from the bottom section of the debutanizer as a feed. This process is used to increase low-octane naphtha fractions to high-octane reformate for use as premium motor fuel blending components or as source of aromatic hydrocarbons.

The *fluid catalytic cracking (FCC)* unit converts gas oils into a range of hydrocarbon products of which gasoline is the most valuable. In any refinery, the amount of low-market-value feedstock available for catalytic cracking is considerable, and a typical FCC unit's ability to produce gasoline from low-market-value feedstock gives the FCC unit a major role in the overall economic performance of a refinery. There are typically two to four feeds that include fresh feeds from the crude and vacuum units, and recycle feed from the main fractionator.

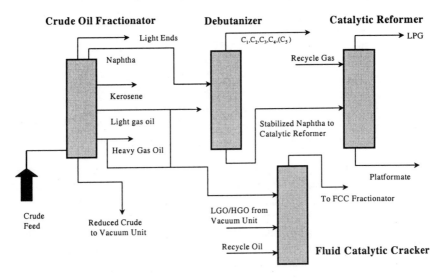

Figure 1.1. Overview of the refinery processes considered for non-linear model-based multivariable control applications

This book is divided into eight chapters. In Chapter 2, a literature overview of the model-based control algorithm is presented. The properties of various algorithms are summarized and the relationship between GMC and other control strategies is discussed in general, and in particular, specific applications of model-based control in industries are highlighted as well as the non-linear model-based control techniques. Non-linear generic model-based control proposed by Lee and Sullivan (1988) is selected for further studies and research work in this book.

Chapter 3 deals with the development of inferential models and their applications in real-time to various refinery processes. The interface of these models to non-linear multivariable control applications is emphasized. The inferential models developed in this chapter are also used in various refinery processes described in other chapters of the thesis. All these models form a closed-loop quality control system with non-linear multivariable control.

Chapters 4 to 7 form the body of the book, which present the various applications of non-linear model-based multivariable control strategies and the development of constrained optimization techniques to a wide variety of refineries processes exhibiting non-linear dynamic responses and process interactions. Chapter 4 uses a steady-state process model in conjunction with approximate dynamics and the resulting non-linear model-based control strategies were then applied to an industrial debutanizer.

In Chapter 5, the Shell heavy oil fractionator problem proposed by Prett and Morari (1987) is studied and a constrained non-linear optimization problem is constructed. An inferential model for top end point is developed and incorporated in the control algorithm. The non-linear control strategies are also applied in real-time to the other crude oil fractionator.

Chapters 6 and 7 address complex control problems on the reactor section of a catalytic reforming process and on the reactor-regenerator section of a fluid catalytic cracking process. Non-linear multivariable control and constrained optimization algorithms are developed and applied in real-time to these processes to demonstrate the success of these strategies as compared to linear control methods in controlling the processes with strong non-linearities and interactions. In Chapter 7, a dynamic parameter update system is also developed and applied to the FCC process.

In Chapter 8, the conclusions from this study and the recommendations and future direction for research are presented. Appendices provide supplementary materials and information and related programme used in the various chapters.

To make the book more clear and compact, while Chapter 2 provides a literature review on model-based control (linear and non-linear), other chapters also present a brief literature review of related techniques and industrial applications.

In particular: Chapter 3 provides a review on inferential models and control; Chapter 4 gives some details on the use of steady-state models in model-based control; Chapter 5 provides related industrial applications and review on the model-based multivariable control application on crude distillation process; Chapter 6 and 7 give a comprehensive review on non-linear model-based control and constrained optimization techniques applied to the complex processes such as the catalytic reforming and the fluid catalytic cracking. Chapter 7 also provides an overview on the process-model mismatch and discusses the problems related to modelling errors and dynamic parameter update system.

CHAPTER 2

MODEL-BASED CONTROL: LITERATURE REVIEW

2.1 Introduction

Model-based control is a generic term for a widely used class of process model-predictive control (MPC) algorithms. Model-predictive control has emerged as a powerful practical control technique during the last decade. Its strength lies in its use of step response data, which are physically intuitive, and that it can handle hard constraints explicitly through on-line optimization. Various MPC techniques such as dynamic matrix control (DMC) (Cutler and Ramaker, 1980), model algorithmic control (MAC) (Rouhani and Mehra, 1982), and internal model control (IMC) (Garcia and Morari, 1982) have demonstrated their effectiveness in industrial applications. As described in chapter one, a process model and a reference trajectory are two of the most essential characteristics of model-based control algorithms such as GMC. Recently, an interesting application based on neural model-predictive control (NMPC) method was proposed by Ishida and Zhan (1995) for the one-step predictive control of MIMO processes.

Process analysis has typically used steady-state mass and energy balance models, whereas process control has used dynamic, linear cause-and-effect models and process optimization has used steady state, linear, economic models. There are several major drawbacks to the use of these widely varying models. Firstly, there is always some degree of mismatch between the various models and therefore, the results from one model may not be applicable to the others. Secondly, there is the overhead in developing, maintaining and correlating several different models to describe the same process. The model used in model-based control is usually obtained by applying fundamental principles, for example: conservation equations and constitutive relations (Lee, 1991). The advantage of using such models in control strategies is that these models can model the non-linear behaviour of the process precisely, have fewer parameters, can be obtained from design studies and the models can also be used for optimization studies.

There is a trade-off between the computational effort involved in solving the model equations and the accuracy of the model prediction. Thus, a reduced model, even to the extent of a linear approximation of a non-linear process model, is often used. The model-based algorithms of internal model control (IMC) (Garcia and Morari, 1982) are based on a linear process model.

The "discrete convolution model" is another form of the process model which has been used in several model-based control algorithms with an advantage that the coefficients of the model can be obtained directly from values of the output responses to step changes in the manipulated variables without assuming a model structure. The model is only applicable to those systems which can be described by a set of linear difference equations (Marchetti *et al.*, 1983).

The desired closed-loop response is represented by the reference trajectory of the model-based control algorithm. The MAC algorithm (Rouhani and Mehra, 1982) and a few other forms of MPC (Ricker, 1985; Caldwell and Dearwater, 1991) included a "reference trajectory", the intent of which was to specify the desired speed of the closed loop response for both setpoint and disturbance outputs. The reference trajectory varies for different algorithms. It should be so specified that the control algorithm can determine values of the manipulated variables that make the closed loop system response follow this reference trajectory. Therefore, the reference trajectory is also dependent on the form of the process model used.

Recently, Duan *et al.* (1997) presented a multivariable weighted predictive control (MV MPC) algorithm and discussed its properties for a closed-loop system. The method is based on polynomial system models. Simulation studies for industrial processes showed that satisfactory closed-loop performance can be achieved by tuning the design parameters in the MPC algorithm.

This chapter presents a review of model-based control algorithms for both the linear and non-linear control systems. In this review, current interest of the processing industry and the latest trends in academic research in model-based control have been highlighted and many industrial applications of model-predictive control have been reported. The need for non-linear model-based control is emphasized and the capabilities of non-linear control techniques to handle the common problems associated with the refinery processes, such as time delays, constraints and model uncertainty are discussed. Finally the stability and robustness issues related to both the linear and non-linear control systems are also discussed. A brief review of generic model control (GMC) for linear and non-linear systems is given and non-linear model-based control is selected for further research, design and development work in this text and for the practical applications in industries.

2.1.1 Industrial Background

The current interest of the processing industry in model-predictive control can be traced back to a set of papers that appeared in late 1970's. Richalet *et al.* (1978) described successful applications of "Model Predictive Heuristic Control" and in 1979 engineers from Shell (Cutler and Ramaker, 1980; Prett and Gillette, 1980) outlined "Dynamics Matrix Control" (DMC) and reported applications to fluid catalytic cracker. In both algorithms an explicit dynamic model of the plant is used to predict the effect of future actions of the manipulated variables on the output. The future moves of the manipulated variables are determined by optimization with the objective of minimizing the predicted error subject to operating constraints. The optimization is repeated at each sampling time based on updated measurement from the plant. Therefore, in the context of model-predictive control the control problem including the relative importance of the different objectives, the constraints, *etc.*, is formulated as a dynamic optimization problem. It constitutes one of the first examples of large-scale dynamic optimization applied routinely in real-time in process industries.

Since the rediscovery of MPC in 1978 and 1979, its popularity in the chemical-process industries has increased steadily. Mehra *et al.* (1982) reviewed a number of applications including a superheater, a steam generator, a wind tunnel, a utility boiler connected to a distillation column and a glass furnace. Shell has applied MPC to many systems such as a fluid catalytic cracking unit (Prett and Gillette, 1979) and a highly non-linear batch reactor (Garcia, 1984).

Several companies (Bailey, DMC, Profimatics, Setpoint) offer MPC software. Cutler and Hawkins (1988) report a complex industrial application to a hydrocracker reactor involving seven independent variables (five manipulated, two disturbance) and four dependent (controlled) variables including a number of constraints. Martin *et al.* (1986) cites seven completed applications and ten under design. They include: fluid catalytic cracker including regenerator loading, reactor severity and differential pressure controls; hydrocracker bed outlet temperature control and weighted average bed temperature profile control; hydrocracker recycle surge drum level control; reformer weighted average inlet temperature profile; analyzer loop control. Caldwell and Martin (1987) have described the latter in more detail. Setpoint (Grosdidier *et al.*, 1993) has applied the MPC technology to various petroleum refining and chemical operations. Many industrial applications have been reported, especially over the last couple of years, and versions of MPC are being marketed by most of the major vendors of process control software.

Ohshima *et al.* (1995) presented model-predictive control algorithm with adaptive disturbance prediction and its application to three distillation columns. Nikravesh *et al.* (1995) applied dynamic matrix control to a diaphragm-type chlorine/caustic electrolysers and compared the results of DMC with PID control.

Ansari *et al.* (1997) implemented DMC on a 120,000-barrels/day crude distillation unit. The DMC application on this unit maximized the yields of high value products, increased the production of LPG and provided better control of the unit. The controllers were implemented in Yokogawa Centum-CS DCS advanced control station (ACS).

Recently Sriniwas and Arkun (1997) presented an interesting case-study where model-predictive control was applied to control a non-linear, open-loop unstable process called the Tennessee Eastman Challenge Process.

2.1.2 Academic Background

In the academic environment, MPC has been applied under controlled conditions to a simple mixing tank and a heat exchanger (Arkun *et al.*, 1986) as well as a coupled distillation column system for the separation of a ternary mixture. Parrish and Brosilow (1985) compared MPC with conventional control schemes on a heat exchanger and an industrial autoclave. Most of the applications reported above are multivariable and involve constraints. It is exactly these types of problems that motivated the development of the MPC control techniques. Garcia *et al.* (1989)

reviewed the technical and economic incentives for the use of MPC and its revolutionary history. There are also several new books on the subject (Prett and Garcia, 1988; Morari and Zafiriou, 1989) and MPC has been the focus of two workshops (McAvoy *et al.*, 1989; Prett *et al.*, 1990). Several MPC algorithms using the general state-space model have also been proposed (Ricker, 1990; Lee *et al.*, 1991). Bitmead *et al.* (1990) presented a lucid and detailed analysis of the basic features inherent in all MPC algorithms from the point of view of linear quadratic regulator and linear quadratic gaussian control theory.

Some researchers have preferred the input-output transfer function model over the state-space model in developing MPC algorithms. Clarke *et al.* (1987a,b) developed what is known as "generalized predictive control" (GPC), based on parametric input-output models and showed its connections to LQ optimal control. Doyle III and Hobgood (1995) presented an approach to approximate feedback linearization which utilizes an approximate input-output normal form model for the process.

Robustness to modelling errors and measurement noise was incorporated into the algorithm through "artificial" disturbance dynamics (Clarke, 1991). Morari and Lee (1991) have developed an MPC approach based on the finite impulse and finite step response models in which the states are simply the appropriate number of past values of the inputs (or change in the inputs). This approach can be extended to processes with pure integrators. Recently an interesting paper of Lundstrom *et al.* (1995) on limitations of dynamic matrix control has been published. The paper separates the DMC algorithm into a predictor and an optimizer that highlights the DMC limitations and suggests how these can be avoided.

2.2 Model-predictive Control (MPC)

The name "model-predictive control" arises from the manner in which the control law is computed. At the present time k the behaviour of the process over a horizon p is considered. Using a model the process response to changes in the manipulated variable is predicted. The moves of the manipulated variables are selected such that the predicted response has certain desirable characteristics. Only the first computed changes in the manipulated variable is implemented. At time k+1 the computation is repeated with the horizon moved by one time interval. The model-predictive algorithms include dynamic matrix control (DMC) (Cutler and Ramaker, 1980), model algorithm control (MAC) (Rouhani and Mehra, 1982) and principal component analysis (PCA) (Maurath *et al.*, 1985b). All of these algorithms are based on a discrete convolution model of the process. Seborg (1987) and Garcia *et al.* (1989) also reviewed these algorithms.

2.2.1 Dynamic Matrix Control (DMC)

DMC belongs to the family of model-predictive controller (MPC) algorithms. The main idea behind these algorithms is to use an explicit model of the plant to predict the open loop future behaviour of the controlled outputs over a finite time horizon.

The predicted behaviour is then used to find sequence of control moves, which minimizes a particular objective function without violating prespecified constraints. The process model used in DMC is a discrete convolution model and uses step response:

$$y(t+1) = y^\circ (t+1) + a_1 \Delta u (t) \tag{2.1}$$

$$y(t+2) = y^\circ (t+2) + a_2 \Delta u (t) + a_1 \Delta u (t+1)$$

$$\vdots$$

$$y(t+k) = y^\circ (t+k) + a_k \Delta u (t) + a_{k-1} \Delta u (t+1) + \ldots + a_1 \Delta u (t+k-1)$$

$$\vdots$$

$$y(t+M) = y^\circ (t+M) + a_M \Delta u (t) + a_{M-1} \Delta u (t+1) + \ldots + a_1 \Delta u (t+M-1)$$

$$\vdots$$

$$y(t+T) = y^\circ (t+T) + a_T \Delta u (t) + a_{T-1} \Delta u (t+1) + \ldots + a_{T-M+1} \Delta u (t+M-1)$$

where $y^\circ(t+k)$ represents the predicted system output at time t+k before implementing any control moves (*i.e.*, no control action changes). $y(t+k)$ is the output with control moves. $\Delta u (t+k)$ represents the change in the input at time t+k (called the control move). a_1 represents the changes observed in the system output at time t+i if a unit step change in the input occured at time t. T is called the prediction horizon, and M≤T is called the control horizon.

The above equations could be rewritten in vector-matrix form as:

$$y = y^\circ + A \, \Delta u \tag{2.2}$$

where the matrix A is called the Dynamic Matrix of the system and is given by:

$$A = \begin{bmatrix} a_1 & 0 & \cdots & 0 \\ \vdots & & & \vdots \\ a_M & a_{M-1} & \cdots & a_1 \\ \vdots & & & \vdots \\ a_T & a_{T-1} & \cdots & a_{T-M+1} \end{bmatrix} \tag{2.3}$$

Define the error prediction vector as:

$$e = y^* - y^\circ \tag{2.4}$$

y^* is the setpoint of the output.

A set of M future input changes Δu (t) ... Δu (t+M -1) is determined by solving the following optimization problem:

$$\text{Minimize } J_{M,T} = (A \, \Delta u - e)^T \, Q^T \, Q \, (A \, \Delta u - e) + r^2 \, \Delta u^T \, \Delta u \qquad (2.5)$$

where Q is weighting matrix. The solution is given by:

$$\Delta u = (A^T \, Q^T \, Q \, A + r^2 \, I)^{-1} \, A^T \, Q^T \, Q \, e \qquad (2.6)$$

The introduction of the tuning parameter r improves the control behaviour when the matrix $A^T \, Q^T \, Q \, A$ is ill-conditioned, ensuring stability.

DMC applies only Δu (t), the first element of the M optimal future control actions $\Delta u(t)$ .... Δu (t + M -1) calculated by the above equation, then updates the error prediction vector at each time instant and repeats the algorithm. The extension of DMC to MIMO system is straightforward (Cutler and Ramaker, 1980; Cutler, 1983).

The robustness of the algorithm can be guaranteed by the tuning parameters M, T, and r, and the sample time interval (Cutler, 1983; Maurath *et al.*, 1985a). Some guidelines can be found in the literature for their selection (Cutler, 1983). DMC would work efficiently for imbalanced (*i.e.*, non-square) systems or systems with unusual behaviour, provided that the step response model can be obtained (Cutler, 1981).

2.2.2 Limitations of Dynamic Matrix Control

Dynamic Matrix Control is based on two assumptions that limit the feedback performance of the algorithm (Lundstrom *et al.*, 1995). The first assumption is that a stable step response model can be used to represent the plant. The second assumption is that the difference between the measured and the predicted output can be modeled as step disturbance acting on the output. These assumptions lead to the following limitations:

1. Good performance may require an excessive number of step response coefficients.
2. Poor performance may be observed for "ramp-like" disturbances acting on the plant outputs. In particular, this occurs for input disturbances for plants with large time constants.
3. Poor robust performance, due to input gain uncertainty (which is always present in practice), may be observed for multivariable plants with strong interactions.

In addition, there is the obvious limitation that the plant has to be stable. Limitations 1 and 2 apply when the plant's open loop time constant is much larger than the desired closed-loop time constant. Limitation 3 is caused by gain uncertainty on the inputs. The DMC algorithm was separated into two parts by

Lundstrom *et al.* (1995), a predictor and an optimizer as shown in Figure 2.1. By splitting up the algorithm in this manner, similarities with state-observer state-feedback controllers become apparent. In fact, Lee *et al.* (1994) show that unconstrained DMC is equivalent to an optimal state observer (Kalman filter) and linear quadratic feedback, using a receding horizon approach and special assumptions about disturbances and measurement noise. Figure 2.2 also shows the objective of the predictor to generate a vector, $y(k+1| k)$, of predicted open loop outputs over a horizon of p future time steps, the prediction horizon. This prediction vector is then used as an input to the optimizer. The algorithm details are given in Lundstrom *et al.* (1995).

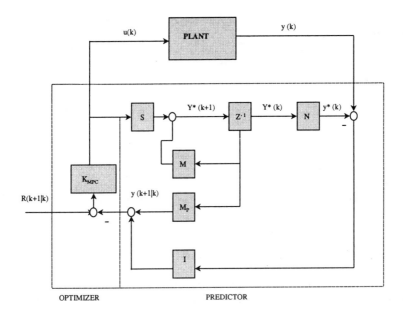

Figure 2.1. DMC Algorithm separated into a Predictor and and Optimizer

There are several variants of DMC: original DMC (Cutler and Ramaker, 1980), DMC with least squares satisfaction of input constraints (Prett and Gillettte, 1980), DMC with constrained linear programming optimization (LDMC) (Morshedi *et al.*, 1985), DMC with constrained quadratic programming optimization (QDMC) (Garcia and Morshedi, 1986). These variants use different optimizers but the predictor is the same for all of them. Both the limiting assumptions, which have been defined above, are implicit in the predictor and will not be avoided by modifying the optimizer, so the results hold for all these algorithms. The results also carry over to the more general case with constraints, since the issue of constraints only affects the optimizer.

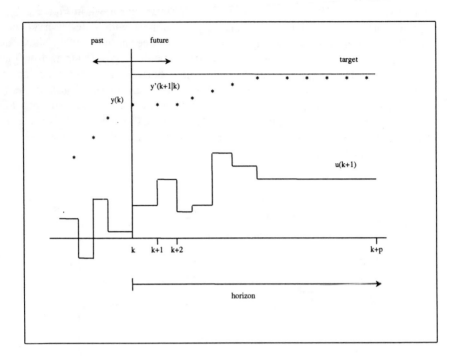

Figure 2.2. Schematic of vector of predicted open-loop outputs over a horizon of p future
time steps

2.2.3 Model Algorithm Control (MAC)

The original idea of model algorithm control (MAC) first appeared as model-
predictive heuristic control (Richalet *et al.,* 1978) without any theoretical analysis,
but as a heuristic design procedure. The basic theoretical properties of stability,
robustness and constraint handling were investigated later after a number of
successful applications on industrial processes (Mehra and Rouhani, 1980; Mehra *et
al.,* 1979, 1980, 1981; Rouhani and Mehra, 1982).

The main strategies of model algorithm control consist of four parts:

1. Instead of the step response, system is represented by an impulse response.
2. Prediction of system behaviour using the process model.
3. A reference trajectory was defined for control system.
4. Developed criterion for optimality leading to an optimal control system
    algorithm.

For a SISO system, the impulse response model is given by:

$$y(t+1) = h^{T} u(t) = h_1 u(t) + h_2 u(t-1) + ... + h_N u(t -N +1) \qquad (2.7)$$

where $y(t+1)$ is the process output at time $t+1$, $h^T \in R^{N+1}$ denotes the process impulse response, $u(t - j)$ is the input at time $(t - j)$ to the process and $u(t) = [u(t) \ldots u(t -N+1)]^T$, N is the dimension (size) of the model. A linear prediction scheme is used in MAC:

$$y_p(t+k) = y_M(t+k) + [y_p(t +k -1) - y_M(t +k -1)]$$
$$= y_p(t+k -1) + h^T[u(t +k -1) - u(t +k -2)] \quad k= 1, \ldots T \qquad (2.8a)$$

$$y_p(t) = y(t) \qquad (2.8b)$$

where $y_p(t+k)$ is the predicted output at time $t+k$, $y_M(t+k)$ is the model output computed by Equation 2.7 and $y(t)$ is the process output measured at time t. The purpose of MAC is to lead the output $y(t)$ along a desired smooth path (called the reference trajectory) to an ultimate setpoint $y*$. The reference trajectory used in MAC is:

$$y_r(t+k) = \alpha^k y_r(t) + (1-\alpha^k) y* \qquad k = 1, \ldots T, \quad |\alpha| < 1 \qquad (2.9a)$$

$$y_r(t) = y(t) \qquad (2.9b)$$

where α is a tuning parameter. T represents the future time horizon over which the process output is to follow the reference trajectory. A set of T future inputs $u(t)$..... $u(t + T-1)$ is then determined by solving an optimization problem:

$$\text{Minimize } J_T(u(t),\ldots u(t+T-1) = \Sigma (y_p(t+k) - y_r(t+k))^2 q_k^2 \qquad (2.10)$$
$$k = 1,\ldots T$$

such that $u(t+k) \in \Omega \qquad k = 0, \ldots T-1$

where q_k^2 are weighting factors. Ω denotes the constraint set for the process inputs. Even though one can use all the elements of the optimum control set $u(t),..u(t+T-1)$ before observing the process outputs and then reapply the whole strategy, at each sampling time interval only the first control action $u(t)$ is applied in MAC. The procedure is then repeated at each sampling interval. The advantage of this implementation is that it keeps the predictions closer to the actual values of the output variable. In addition, if the input set $u(t),...u(t+T-1)$ is free of constraints, then the length T of the prediction horizon does not affect the optimum value of the first input $u(t)$, i.e., the solution of Equation 2.10 has the same $u(t)$ for different values of T. The extension of MAC to MIMO system is straightforward (Mehra et al., 1979).

Generalized approach for model algorithm control (GAMAC), developed by Hmood and Prasad (1987), has the same basic structure as MAC. It consists of a reference model, a regulator, a feedforward and a feedback controller. The authors also developed a method to treat systems with nonminimum phase response, which is similar to a pole replacement technique. Simulation results of several SISO minimum phase and non-minimum phase systems were presented by the authors, but application to real process has not been reported.

Cheng (1989a,b) constructed an algorithm based on a discrete linear state space model with similar prediction, reference trajectory and optimality criterion as MAC. However, the objective function of the optimization is a sum of squared terms taken from the present time to infinity. This algorithm was referred to as linear quadratic MAC because the final optimization form is the same as that of a standard linear quadratic feedback control method. This method was applied to a simulated multiple-effect evaporator with and without process/model mismatch.

2.2.4 Difference between DMC and MAC

1. Instead of the step response model used in DMC, MAC uses an impulse response model. However, the coefficient of these two models have the following relationship (Marchetti *et al.*, 1983)

$$a_l = \Sigma\, h_j j = 1,...I \qquad\qquad (2.11)$$

   If the input is penalized in the quadratic objective, then the controller does not remove the offset. This can be corrected by a static offset compensator (Garcia and Morari, 1982). If the input is not penalized then extremely awkward procedures are necessary to treat non-minimum phase systems (Mehra and Rouhani, 1980).
2. In model algorithm control, the control horizon is equal to the prediction horizon.

$$M = T \qquad\qquad (2.12)$$

3. The reference trajectory of DMC is equal to a constant setpoint while the reference trajectory of MAC is defined through a tuning parameter α. Increasing the tuning parameter avoids excessive control movements. However, a tuning parameter r is introduced in the objective functions of DMC to suppress the control action.

For SISO systems, Ogunnaike and Adewale (1986) modified the DMC algorithm so as to account for time variations in the system deadtime and steady-state gain by a simple operation that required no on-line matrix inversions. Its simulated applications to a distillation column and a furnace was presented.

2.2.5 Principal Component Analysis (PCA)

A new predictive control algorithm approach to the optimization problem was designed by Maurath *et al.* (1985b, 1988) and termed "principal component analysis" (PCA). That is, when the matrix $A^T Q^T Q A$ is ill conditioned, singular value decomposition (SVD) is used to decompose the matrix QA as:

$$Q A = U \Sigma V^{T}, \Sigma = \begin{bmatrix} S & 0 \\ 0 & 0 \end{bmatrix} \tag{2.13}$$

where U, V are orthogonal matrices, S is a diagonal matrix with elements that are the non-zero singular values of QA. Then one can re-write the objective function of the optimization problem with r=0 as:

$$J(\Delta u) = \bullet\, Q A \, \Delta u - Q e \bullet_{2}^{2} \tag{2.14}$$

or

$$J(w) = \bullet\, \Sigma w - g \bullet_{2}^{2} \tag{2.15}$$

which could be solved easily without numerical difficulties. Here $w = V^{T}\Delta u$ and $g = U^{T} Qe$. The elements of vector w are the principal components of the solution to the least square problem. Moreover, the singular value decomposition provides some information on the system performance and the size of the control moves. This information can be used to balance the considerations of system performance and robustness by selecting the appropriate principal components.

The regulatory and servo response of PCA were demonstrated in designing and evaluating satisfactory controllers for two simulated multivariable distillation columns (Maurath et al., 1985b, 1988). Callaghan and Lee (1986, 1988) used PCA for a simulation of a grinding circuit and to an experimental heat exchange network. The heat exchange network exhibited non-linear behaviour which has different open-loop responses for a unit step-down and step-up change of the input. Callaghan and Lee (1988) used an approximate model obtained from an arithmetic average of the step-up and step-down responses, which gave good results.

2.3 Internal Model Control (IMC)

2.3.1 IMC Theoretical Background

The motivation behind the development of internal model control (IMC) was to combine the advantages of the different unconstrained MPC schemes and to avoid their disadvantages: easy on-line tuning via adjustment of physically meaningful parameters and without any concern about closed-loop stability; good performance without inter-sample rippling; ability to cope with inputs other than steps.

Internal model control (Garcia and Morari, 1982) puts the model-predictive control algorithms such as MAC and DMC into a theoretical framework. IMC was designed in the Z-transform domain or in the Laplace transform domain and is also based on a linear model of the process. The structure of IMC with a filter in the Z-transform domain is shown in Figure 2.3.

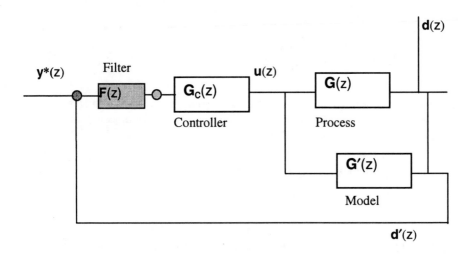

Figure 2.3. The internal model control (IMC) Structure

The IMC structure can be converted into a classical feedback control structure (Garcia and Morari, 1982). Three important properties have been proven for SISO system under the condition that F(z) =1 (Garcia and Morai, 1982):

When the model is exact, stability of both the controller and plant is sufficient for overall closed-loop system stability.

Under the assumption that $G'(z) = G(z)$ and that $G(z)$ is stable, the sum of the squared errors is minimized for both the regulator and the servo-controller when:

$$G_C(z) = 1/G_-(z) \tag{2.16}$$

where $G'(z) = G'_+(z) \, G'_-(z)$, $G'_+(z)$ contains all time delays and all zeros outside the unit circle.

A controller G_C which satisfies $G_C(1) = G'(1)^{-1}$ yields zero offset. Therefore the critical point in IMC is how to approximate $H^{-1}(z)$, where $G'(z) = z^{-\tau-1} H(z)$ and τ is the time delay of the process.

Garcia and Morari (1982) investigated the robust and servo-behaviour of IMC and compared its behaviour and structural resemblance with other algorithms such as Inferential Smith-Predictor (Brosilow, 1979) and Linear Quadratic Optimal Control.

The implementation of the IMC algorithm is almost the same as MPC. At each time instant t, the solution of the following optimization problem gives the control law:

$$\text{Min} \sum_{k=1}^{T} \gamma_k^2 \, [y_p(t+\tau+k) - y_d(t+\tau+k)]^2 + \beta_k^2 \, u(t+k-1)^2 \qquad (2.17)$$

w.r.t $\qquad\qquad$ u(t) ... u(t+M-1)

such that $y_p(t+\tau+k) = y_m(t+\tau+k) + d(t+\tau+k)$
$\qquad\qquad = h_1 u(t+k-1) + h_2 u(t+k-2). \ldots + h_N u(t+k-N) + d(t+\tau+k)$
$\qquad\qquad u(t+M-1) = u(t+M) = \ldots = u(t+T-1)$
$\qquad\qquad \beta_k = 0, \quad k > M$

where T is the optimization horizon, $y_d(t+\tau+k)$ is the desired trajectory, γ_k^2 is a time-varying weight on the output error, β_k^2 is a time-varying weight on the input and M is the input suppression parameter which specifies the number of time intervals into the future during which u(t+k) is allowed to vary. $y_p(t+\tau+k)$ is the predictive output, $y_m(t+\tau+k)$ is the output of the model, $d(t+\tau+k)$ is the predicted disturbance.
Let:

$$d(t+\tau+k) = d(t) = y(t) - y_m(t) \qquad\qquad k = 1, \ldots T \qquad (2.18)$$

then M, T, β_k, γ_k are tuning parameters of the algorithm which affect the stability and robustness of the controller. Only the first input move is implemented, just as in MAC and DMC.

2.3.2 Comparison of IMC with DMC and MAC

1. The structure inherent in all MPC schemes, referred to as IMC structure, corresponds to a very convenient way of parametrizing all stabilizing controllers for an open-loop stable plant. This makes the structure very useful for design: a stable MP controller implies a stable closed-loop system and *vice versa*.
2. IMC reduces the number of adjustable parameters of DMC and MAC to a minimum and expands the role of the MAC filter. The controller can also be tailored to accommodate inputs different from steps in an optimal manner. For low-order models the IMC design procedure generates PID controllers with one adjustable parameter.
3. The advantage of unconstrained MPC (with the possible exception of IMC) over other LQ techniques are still to be demonstrated with more than case-studies. Contrary to other techniques, however, the MPC ideas can be extended smoothly to generate non-linear time varying controllers for linear systems with constraints.

2.3.3 Extensions and Variants of IMC

Extension of IMC's theory and implementation on MIMO system has been developed (Garcia and Morari, 1985a, 1985b). The three basic properties remain true for MIMO systems. Difficulties arose from the different time delays of the outputs to inputs. The important role of the filter ensuring robustness was also examined (Garcia and Morari, 1985a). Implementation strategies similar to those for SISO systems were derived for MIMO systems. The effects of the tuning parameters and a suitable tuning procedure were also presented (Garcia and Morari, 1985b). For most models used to describe the dynamics of Chemical Process Systems, Rivera et al. (1986) related a PID controller's tuning parameter to IMC, thus providing a new tuning procedure for a PID controller. These results were developed in the absence of non-linearities, constraints and multivariable interactions.

Economou and Morari (1986) also used IMC to design a multiloop controller, considering the interaction as a special kind of disturbance (modelling error), which could be suppressed by the filters. In order to evaluate different multiloop structures, an IMC interaction measure was introduced.

To account for constraints in processes, Ricker (1985) used Quadratic Programming problem to solve the corresponding constrained optimization problem in a IMC structure. This is very similar to quadratic dynamic matrix control (QDMC).   Application to a simulated evaporator was performed. For a SISO system Rotea and Marchetti (1987) used a linear quadratic regulator for optimization and the constraints could be handled by a single tuning parameter. Holt and Morari (1985a, 1985b) and Morari et al. (1987) used the IMC structure to study the effect of deadtimes and right-half-plane (RHP) zeros on the control performance. Zarifiou and Morari (1986, 1987,1988), and Morari and Zarifiou (1989) studied how to design the filter in IMC to ensure the robustness of the controller by μ-optimal control theories.

The relationship between IMC and MIMO linear quadratic optimal control was studied by Harris and McGregor (1987), and Grimble et al. (1989). It was shown that they could be equivalent under some conditions. The construction of the approximate model for system where the exact model inverse is impossible was studied by Kozub et al., (1989). The approximate model inverse is constructed by a matrix spectral factorization approach. An approach to design IMC controllers for unstable systems was developed by Chiu and Arkun (1990) using coprime fraction representations. The applications of IMC to laboratory experiments (coupled distillation column) and pilot plants (heat exchanger, packed bed reactor, stirred water tank) were reported by Arkun et al. (1986), Levien and Morari (1987), and Kozub et al. (1987).

Linear model-dependent control (MDC), based on IMC theory, was designed for SISO systems (Papadoulis and Svoronos, 1987), which could handle open-loop instability, nonminimum phase response and time delays by defining a generalized input. This generalized input is a function of the control variable and the output variable. It makes the open-loop zeros close to the origin by introducing a parameter. Then IMC is used for the new system to obtain the control law, which can be mapped into the original control variable. The stability and zero offset property of MDC can be obtained by tuning two parameters (one for open-loop zeros and one for the filter).

Simplified model-predictive control (SMPC) (Arulalan and Deshpande, 1986, 1987) is another algorithm based on IMC for open-loop stable systems (both SISO and MIMO). Using an impulse response model for system representation and by specifying the closed-loop response, the procedure for approximating the inverse of the process as outlined in IMC could be simplified. A parameter was introduced to speed up the system response. A filter could be added to ensure the algorithm's robustness. Application to a binary distillation column was shown by simulation.

Linear inferential control (LIC) (Joseph and Brosilow, 1978; Brosilow, 1979) and non-linear inferential control (NLIC) (Parrish and Brosilow, 1986, 1988) have a similar structure as IMC except that an observer is incorporated for unmeasured state variables. For non-linear systems, as in IMC, the inverse of a non-linear operator (the process model) has to be calculated. Recently Häggblom (1996) presented a combined application of IMC with inferential control. For comparison, the author performed some experiments with IMC only. However, the control system is realized as a combined internal model and inferential control (CIMIC) system.

2.4 Stability and Robustness of Linear MPC

The design of a model-based control system depends on whether the model can effectively characterize the process. Since plant/model mismatch that includes modelling error, variation of the process parameters, *etc.*, is almost inevitable in practical applications, the mismatch should be considered for control system design. Robust control is able to deal with the problems of plant/model mismatch. Robust stability and robust performance have been extensively studied. Morari and Doyle (1986) showed how robust control can address real-life problems and deal with several types of model uncertainty. Robust performance in linear control is usually achieved by minimizing the error for the worst-case scenario. Since MPC also minimizes the error by solving a constrained "min max" problem. Efforts in this direction (Campo and Morari, 1987; Cuthrell *et al.,* 1990; Beigler 1990; Palazoglu *et al.,* 1989) concentrated on ways to solve the complex optimization problems on line.

Garcia and Morari (1982) proved four theorems about the stability of the IMC controllers described in Section 2.3:

1. For $\gamma_i \neq 0$, $\beta_i \neq 0$, selecting $M = T > N$ yields the model inverse control G_C (z) = H^{-1} (z).
2. Assume that the system has a discrete monotonic step response and that $\gamma_i = 1$ $\beta_I = 0$, T= N. For M chosen sufficiently small the controller is stable.

3. There exists a finite $\beta^* > 0$ such that for $\beta_i \geq \beta^*$ ($i = 1 \ldots M$), the controller is stable for all $M \geq 1$, $T \geq 1$ and $\gamma_i \geq 0$.
4. Assume $\gamma_i = 1$, $\beta_i = 0$. Then for sufficiently small M and sufficiently large $T > N + M - 1$ the controller is stable.

According to these theorems, tuning procedures for the sampling time Ts, the input suppression parameter M, the input penalty parameter β_i, the output penalty parameter γ_i and the optimization horizon T were designed for minimum phase systems and nonminimum phase systems separately (Garcia and Morari, 1982).

It was pointed out (Garcia and Morari, 1982) that with specific choices of tuning parameters, MAC was a special case of IMC where an exponential filter was specified. However, for systems with nonminimum phase response, IMC's strategy was far better than MAC with respect to the computational efforts. It was also claimed that DMC was equivalent to IMC with $T = N$ and $M < N$ and with minor modifications of the objective function.

Mehra *et al.* (1980) demonstrated MAC's servo behaviour and that if the corresponding auto-regression model of the optimal input sequence u*(t) is stable, then the closed loop output is stable if perfect model identification occurs. The input-output stability following a change of setpoint is totally dependent on the process being minimum phase.

The robustness of MAC has been discussed under several types of plant/model mismatch (e.g., SISO with additive noise, constant gain mismatch, constant matrix mismatch) (Mehra *et al.*, 1979; Rouhani and Mehra, 1982). It was shown that the robustness could be improved through tuning the reference trajectory parameter. Garcia and Morari (1982) also reached the same conclusion. Mehra *et al.* (1979) also stated "The particularly robust character of MAC stems, at least partially, from the redundancy of the input-output representation (impulse response representation)." The compensation of deadtime is inherent in the structure of MAC. Constraint handling is directly implemented in the optimization phase, and it has been proven that the stability of the system with inputs free of constraints implies the stability of the input constrained system (Mehra *et al.*, 1980; Rouhani and Mehra, 1982).

To control a system with nonminimum phase response, one must construct the process model so that the closed loop system is stable. Mehra *et al.* (1980, 1981) selected an optimum model based on minimizing the Euclidean distance of the output to the reference trajectory, which requires the off-line solution of a Riccati matrix equation. Mehra *et al.* (1981) has reported that MAC has been applied successfully to superheater control and on-line control of a steam generator. A list of other applications have been documented by Garcia *et al.* (1989).

To deal with the effect of the process/model mismatch, Grimm *et al.* (1989) derived a robustness criterion for PCA, which is a sufficient condition to ensure robustness for a given mismatch. The criterion quantitatively explored the number of the singular values used in the control law should be limited to ensure the robustness while as many as possible singular values would be used in order to achieve good performance.

Recently Noh *et al.* (1996) proposed a modified robust generalized predictive control (GPC) to improve the possibility of the existence of a solution for robust constrained GPC. These authors also compared the simulation results with conventional GPC.

Jun *et al.* (1996) presented Horizon Predictive Control (HPC) application to a binary distillation column. The work describes a MATLAB®-based computer aided design tool which accomplishes integrated system identification and robustness analysis for HPC. The model-predictive control algorithm was implemented in Honeywell TDC-3000 Distributed Control System.

2.5 Non-linear Model-based Control (NMBC)

2.5.1 Introduction

It is well understood that one of the characteristics of chemical processes that present a challenging control problem, is the inherent non-linearity of the process. In spite of this knowledge, chemical processes have been traditionally controlled using linear systems analysis and design tools. The use of a linear system technique is quite limiting if the process is highly non-linear. Although the computational demands for linear system simulation and implementations are small as compared to non-linear simulation, the recent advances in control system software and hardware have made the practical application of non-linear model-based control system techniques much easier.

A number of review articles from the early 1980's mentioned non-linear system techniques very briefly. Ray (1983) surveyed the field of multivariable process control, with non-linear system playing an insignificant role. During 1980's, the only published literature focussing on non-linearities was from Ray (1982), Shinner (1986) and Morshedi (1986). Lee and Sullivan (1988) developed a procedure based on reference system synthesis (RSS) techniques which uses a model of the process and a desired process response to determine a feedback control law. They called this method a generic model control (GMC). Henson and Seborg (1990) have shown that GMC and internal decoupling are implicitly based on differential geometric concepts. Lee *et al.* (1989) and Newell and Lee (1989) presented a complete case-study of GMC of a single effect evaporator system. GMC was compared with DMC and showed excellent dynamic behaviour. The details of GMC techniques and its comparison with other methods are given in Section 2.6. Bequette (1991) reviewed much of the process control literature on model-based control as well as other control approaches for non-linear plants. The exploration of new feedback strategies and model forms continues. Sistu *et al.* (1993) further examined the computational issues for non-linear model-based control. Gattu and Zafiriou (1992) extended DMC for non-linear systems and used a constant disturbance model and state estimation for feedback control. De-Oliveria and Biegler (1995) provided a discussion of the Newton-type controller for constrained non-linear system. Hernandez and Arkun (1993) developed a small predictive control strategy for non-linear systems based on polynomial ARMA models.

Neural-network-model-predictive control (NNMPC) is another typical and straightforward application to non-linear control. Ishida and Zhan (1995) proposed a NMPC method for the one-step predictive control of MIMO processes, which may be classified as an IMC technique, based on the steepest descent optimization algorithm. Mills *et al.* (1995) also discussed similar approaches for adaptive model-based control using neural networks. Spangler (1994) reported that non-linear optimization codes (e.g., SQP) can be integrated with a virtual analyzer neural network model to compute setpoints for process inputs. Recently, Keeler *et al.* (1996) has developed "The Process Perfecter" software based on neural network modelling techniques, which enables closed-loop control and optimization of non-linear processes

Rawlings *et al.* (1994) stated that many of the recent advances in non-linear control theory rely on feedback linearization. A general approach for model-based control of batch crystallizers has been proposed by Miller and Rawlings (1994). They found that the necessity for optimal control varied greatly depending upon the kinetics of growth and nucleation.

Ansari and Tadé (1998) developed and applied non-linear model-based multivariable control algorithm and strategies in real-time to an industrial debutanizer to control the top and bottom product qualities. The control performance was compared with the traditional PID-type control system. In the non-linear control strategies, the authors also incorporated the inferential models. The detail of this work is given in Chapter 4.

Ansari and Tadé (1998) also developed constrained non-linear multivariable control and optimization strategy for handling constraints and applied in real-time to a catalytic reforming reactor-section. The non-linear model-based multivariable controller controls the weighted average inlet temperature (WAIT), while respecting the heater tube temperature constraints. WAIT is set by an octane inferential model forming a closed-loop WAIT/octane quality control. The detail of this work is given in Chapter 6.

2.5.2 Non-linear Model-based Control Architecture

The non-linear model-based multivariable control developed in this text is the core technology of model-based control. Figure 2.4 shows the non-linear model-based control architecture. Integrated within the architecture are: NMBC-Model, for extraction of models from process data; NMBC-Build, for controller configuration, simulation and tuning; NMBC-Connect, for interfacing the controller to the process; and NMBC-Display, a man machine display.

2.5.2.1 NMBC-Model
Development of a mathematical representation of the process is an important part of any model-based control system design. Process modelling is by no means trivial, and the model complexity should be a function of the planned use of the model. Denn (1986) has discussed process modelling emphasizing the computational methods. A process model is a functional relationship among variables that explains the cause and effect relationships between inputs and outputs, respectively. Process

identification involves determining such a model and is required for implementing the model-based multivariable control.

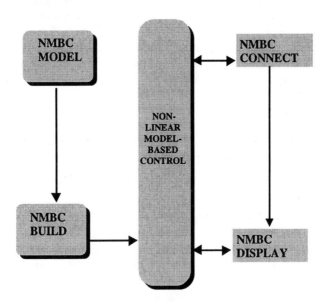

Figure 2.4. Non-linear model-based control architecture

Another method for practical process identification is the black-box approach, where models are obtained exclusively from plant testing data. These models do not consider the non-linearities of the process but provide a description of the dynamic relationship between inputs and outputs and are adequate for satisfactory control. Recently, a new modelling technique, neural networks (Spangler, 1994 and Keeler *et al.,* 1996) has shown the capability to efficiently represent non-linear functional relationships between variables. This technique assumes that the models will approximate the qualitative dynamics of the process adequately, but may not explicitly take into account the qualitative dynamics of the particular system under study.

It is the non-linearities of the process model that provide the improved performance, therefore any controller structure developed must preserve these non-linearities. The generic model control (GMC) structure of Lee and Sullivan (1988) permits the direct use of non-linear dynamic models and, therefore, provides the basic structure of the model-based controller. Since the GMC imbeds a mechanistic process model within the control strategy, this model could also be used for on-line optimization of the process.

Recently, Read and Ray (1998) proposed a novel identification and controller design method for non-linear systems and applied it to a CSTR as plant to be controlled. The identification algorithm is based upon normal form theory, a method borrowed from the dynamic analysis areas. The goal of normal form theory is to drive simple non-linear systems with polynomial vector fields. These polynomial models derived around singularities (points in the parameter space, which organize certain dynamic behaviour) yield simple dynamic models. Knowing the qualitative dynamics, which the singularity contains and the structure of the normal form, the relationship between the structure of a simple dynamic model and its dynamics is solved.

In Chapter 4, a steady-state process model for debutanizer was developed by selecting the Smith-Brinkley's modelling approach (Perry and Green, 1984). The model was used in non-linear multivariable control system with the approximate dynamics obtained by process identification test.

Ansari and Tadé (1998) developed a non-linear process model of a catalytic reforming process and used the model in non-linear multivariable control applications to provide target values for reactor inlet temperatures. The idea was to use a short-cut approach to develop a non-linear process model without sacrificing the non-linearities of the process. The detail of the modelling work can be found in Chapter 6.

2.5.2.2 NMBC-Model Parameter Update

A parameter update system consists of determining the values of the model parameters that produce the best agreement between the model and the real process. It is unlikely that a mechanistic model with one set of parameter values can accurately represent a real process over changing long-term operating conditions. These changes can be caused by fouling of heat exchangers, catalyst deactivation, *etc.*, so the ability to update the model parameters to reflect these changing conditions will improve the accuracy of the model's predictions.

Jang *et al.* (1987) and Chen and Joseph (1987) used an estimation phase, based on optimization to update model parameters and unmeasured state variables. Sistu and Bequette (1990) have shown the advantage of using multi-rate sampling of temperature and composition measurements for rapid disturbance rejection in an endothermic reactor. Rawlings *et al.* (1989) and Meadows and Rawlings (1990) estimated parameters on-line in a semi-batch reactor, using sequential quadratic programming for optimization. The objective of the optimization-based approach is to minimize a measure of model-plant mismatch over an estimation horizon, by using process parameters, unmeasured state variables and loads as decision variables.

Lee *et al.* (1989) presented a process-model mismatch compensation algorithm for model-based control. This algorithm compensated for modelling errors and updated the model parameters at steady state. Signal and Lee (1992) have adopted a more practical approach by proposing the generic model adaptive techniques. Chapter 7 gives a comprehensive review on the dynamic parameter update system, providing an algorithm to update the parameters in real-time application on a fluid catalytic cracking process.

2.5.2.3 NMBC- Controller Configuration/Simulation

Controller simulation and configuration is an important step in the development of any controller design (linear or non-linear). The configuration step defines the basic control strategy and, simultaneously, specifies host database requirements. The simulation step is interactive. It is used to validate controller design and select initial tuning parameters. Once model identification, controller configuration and simulation are completed, data files are transferred to the target computer for the actual application. The details on controller configuration, simulation and on-line tuning are given in Chapters 4 to 7 covering real-time implementation.

2.5.3 Non-linear Control System Techniques

A wide range of methods and techniques has been applied to control the non-linear processes and a brief review is given in the introduction Section 2.5.1. In this section, some specific techniques to control the non-linear processes are described.

Kantor (1989) has developed a *sliding-mode technique* for controller surge vessels. The controller operates by switching between two non-linear feedback rules. The idea is derived from an approach used by McDonald *et al.* (1986) to determine the outlet flow from a vessel that results in the smallest maximum rate of change in outlet flow, while satisfying level constraints. Suarez-Cortez *et al.* (1989) applied sliding mode control to a continuous fermentation process.

Lu and Holt (1990) used a *min-max optimization procedure* to find the optimal control for a discrete, non-linear process. This formulation found the best control for the worst-case plant, based on parameter bounds. The controller was implemented as table look-up controller. A partial solution to the Shell Control Problem (Prett *et al.,* 1989), a distillation column model with manipulated variable uncertainty and constraints, was presented.

Agarwal and Seborg (1987a) have extended SISO *linear self-tuning control strategies to discrete non-linear SISO systems*. They presented simulation results for a CSTR model and showed superior performance over linear adaptive and other non-linear adaptive strategies. Agarwal and Seborg (1987b) extended their previous results to the multivariable case and applied this approach to a distillation model from Wong (1985). Wong and Seborg (1986a,b) developed a low order non-linear model of a distillation column by finding correlation process gains and time constants as a function of the inputs and states.

Manousiouthakis (1990) has used a *game theoretic approach* to robust controller synthesis that is applicable to linear and non-linear systems alike. The idea was to find the controller that achieves the best bound on the worst close-loop performance. The capabilities of these techniques and other methods described in Section 2.5.1, to handle the common problems associated with refinery processes, such as time-delays, constraints and model uncertainty are discussed in the following subsections.

Luyben (1968) has developed *non-linear feedforward control strategies* for a number of CSTR configurations. Non-linear feedforward control was compared with linear feedforward and open-loop control strategies, and performed well even with model uncertainty. For a number of configurations the strategy has an explicit solution, but for many configurations the non-linear feedforward solution has an implicit solution. Rovaglio *et al.* (1990) used a complex, non-linear distillation model on-line to determine the manipulated variable value that best rejects measured feed composition disturbances. Feedforward and feedback linearization was also considered by Calvet and Arkun (1988a,b) using the internal linearization approach. Daouditis and Kravaris (1989) extended GLC to incorporate feedforward/feedback strategies for SISO non-linear systems. Lots of work has also been done for MIMO non-linear systems by various authors.

Deadtime (or *time delays*) frequently occurs in chemical engineering processes because of transport delays of fluids or measurement delays. In practice, compensation of deadtimes in control systems is very important. The Smith predictor (Smith, 1957) is a well known deadtime compensation technique. It utilizes a process model to add feedback around the conventional controller, which eliminates the deadtime from the characteristic equation of the closed-loop system, and converts the control problem with deadtime to one without deadtime.

Kravaris and Wright (1989) developed a technique for deadtime compensation for SISO systems that have deadtime on the manipulated variable. The process must be open-loop stable and the "deadtime-free" part of the process must have a stable inverse. Kravaris (1988) showed an input-output linearization of non-linear systems that is the analog of pole-zero cancellation in linear systems. Kravaris and Daouditis (1990) also extended the linear system idea of placing controller poles at the reflection of the process right-half-plane zeros, to non-linear systems. The performance of this approach was illustrated on an isothermal series/parallel reaction system.

Lee et al. (1990) presented a new method for multivariable dead-time compensation for a wide class of chemical engineering process control problems. Bequette (1990b) and Sistu and Bequette (1990) included deadtime compensation in their NLPC strategy, and showed the effect of time-delay uncertainties. Li and Beigler (1989) have incorporated deadtime in their multi-step approach to non-linear control and used a CSTR example with deadtime between state variables.

The ability to *handle constraints*, which generally are non-linear functions of input, state and output variables, is an important feature for industrial control algorithms when they are applied to real processes. Constraint handling procedure developed by Calvet and Arkun (1988) implicitly handles manipulated variables constraints. An input mapping transformation was used to back calculate the transformed manipulated variable when the physical manipulated variable is constrained. The limitation of this approach is that the controller does not explicitly accounts for the effect of the manipulated variable constraint.

Brown et al. (1990) presented a constrained non-linear multivariable control algorithm where constraint variables can be handled to the constraint limits as well to the rate of approach to the constraint limit.

The method of constrained non-linear multivariable control was applied in real-time to more complex processes such as catalytic reforming and catalytic cracking in Chapters 6 and 7. The results of non-linear multivariable controllers were compared to linear multivariable controllers such as dynamic matrix control. The optimization problem was solved at every time step using the successive quadratic programming (SQP) algorithm, in which the search direction is the solution to a quadratic programming problem.

2.5.4 Non-linear Programming Methods

The constrained non-linear programming methods such as successive quadratic programming (SQP) or generalized reduced gradient (GRG), employed these days are much more efficient than the old methods and technology of 1960's and 1970's. These programming methods can be efficiently coupled to differential and algebraic equations (DAE) process models. Also the modern simulation tools such as NPSOL (Gill *et al.,* 1986), MINOS (Murtagh and Saunders, 1987), NOVA (ABB Inc.), RT-Opt. (AspenTech.) have made optimization-based control schemes inexpensive and routine for small problems. Moreover, software for the creation of non-linear dynamic models is becoming increasingly more available and widely used.

Several methods can be used to handle ordinary differential equations equality constraints with a constrained non-linear optimization programme. Of particular mention are *sequential solution*, iteratively solving the ODEs as an "inner loop" to evaluate the objective function, *simultaneous solution*, transforming the ODEs to algebraic equations which are solved as non-linear equality constraints in the optimization, *intermediate solution*, transforming the ODEs to algebraic equations which are solved as an "inner loop" to evaluate the objective function, and *linear approximation*, either by a single linearization over the prediction horizon, or a linearization at a number of time steps in the prediction horizon.

Morshedi (1986) presented an extension to QDMC called universal dynamic matrix control (UDMC). The modelling equations were integrated using non-linear ODE solver. An efficient approach developed by Morshedi *et al.* (1986) to compute the Jacobian of the states with respect to the manipulated variables.

Balchen *et al.* (1989a) applied a general non-linear predictive technique to an electrometallurgical process, which has a long sample time. Balchen *et al.* (1989b) formulated a predictive optimization problem with constraints on the state and control variables, but indicated that the solution was very time consuming.

Another approach was to reduce the ODEs to algebraic equations by using a weighted residual technique. The equations then solved as equality constraints in a non-linear programme. Cuthrell and Biegler (1987, 1989) have used successive quadratic programming (SQP) to solve optimal control problems. The optimal control profile was calculated and implemented, without compensation for model uncertainty or disturbances.

Patwardhan *et al.* (1990) analyzed the effect of parameter uncertainty and manipulated variable constraints using NLPC of an exothermic CSTR, and compared their results with internal linearization and PI control.

The intermediate solution method utilizes the sequential optimization and simulation approach to solve the algebraic equations resulting from a weighted residual technique. This approach was used by Bequette (1991) and Sistu and Bequette (1990). Bequette (1990) showed the effect of manipulated variable velocity constraints on the control of a biochemical reactor. Bequette (1991) has found that, for SISO systems, the control horizon can generally be set to one time step with the prediction horizon varied for performance and robustness. These results are consistent with linear MPC (Maurath *et al.,* 1988).

In recent years, many authors presented various techniques and methods to solve the non-linear control problems and these methods were discussed in the introduction Section 2.5.1.

2.6 Generic Model Control (GMC)

2.6.1 Introduction

Lee and Sullivan (1988) presented a framework for process controllers that allow for the inclusion of a non-linear mathematical model to approximate the plant behaviour. The so-called generic model control (GMC) formulation of the control law allows process models linear and non-linear, to be embedded directly into the control structure providing the potential for better process control due to more accurate process model being incorporated in the control structure. In this section, a brief review of GMC system (linear and non-linear) is given and then it is compared with internal model control and model-predictive control.

Define a required trajectory for the derivative of the output variables:

$$\dot{y} = k_1 (y^* - y) + k_2 \int (y^* - y) dt \qquad (2.19)$$

where y^* is the setpoint of y, k_1 and k_2 are constant diagonal matrices which can be used to shape any desired trajectory y(t). The reference trajectory of Equation 2.19 has a strong appeal and resembles PI control.

If an approximate process model can be defined as:

$$\dot{x} = f(x, u, d, t) \qquad (2.20)$$

$$y = g(x) \qquad (2.21)$$

where x is the state vector of dimension n, u is the input vector of dimension m, y is the output vector of dimension p, and d is the measured disturbance vector of dimension 1, f and g are vectors with appropriate dimensions, and generally their elements are non-linear functions and t is the time. Substituting Equation 2.20 into Equation 2.19, the control law is given as the solution of:

$$\dot{y} = \frac{\partial g}{\partial x} f(x, u, d, t) \qquad (2.22)$$

It is desirable to have the system operate within the feasible region, such that for the p constraints:

$$C_L \leq C(\mathbf{y}, \mathbf{x}, \mathbf{u}, t) \leq C_U \qquad (2.23)$$

where $\mathbf{C}$, $\mathbf{C}_L$ and $\mathbf{C}_U$ are vectors of dimension p, and each element of $\mathbf{C}$ is generally a non-linear function of $\mathbf{y}$, $\mathbf{x}$, $\mathbf{u}$ and t. These constraints are usually derived from economic and process considerations. The $\mathbf{C}_L$ and $\mathbf{C}_U$ are the lower and upper bounds of vector $\mathbf{C}$.

Both input constraints and input movement constraints are also defined for the n controls, such that:

$$\mathbf{u}_L \leq \mathbf{u} \leq \mathbf{u}_U \qquad (2.24)$$

where $\mathbf{u}$ is an input vector and $\mathbf{u}_L$ and $\mathbf{u}_U$ are the lower and upper bounds of $\mathbf{u}$.

$$\Delta\mathbf{u}_L \leq \mathbf{u}(t + \Delta t) - \mathbf{u}(t) \leq \Delta\mathbf{u}_U \qquad (2.25)$$

where $\Delta\mathbf{u}_L$ and $\Delta\mathbf{u}_U$ are the lower and upper bounds on the input movement vector $\Delta\mathbf{u}$.

A set of performance slack variables can also be incorporated into the GMC performance curves to denote the system's efficiency in terms of setpoint tracking. If the two m-dimensional performance slack variable vectors, λ^-_p and λ^+_p, are defined to express the system's negative offset and positive offset from the pre-specified response trajectory, the GMC control law given by Equation 2.22 can be written for the system performance as:

$$\frac{\partial \mathbf{g}}{\partial \mathbf{x}} \, \mathbf{f}(\mathbf{x}, \mathbf{u}, \mathbf{d}, t) + \lambda^+_p - \lambda^-_p = \mathbf{k}_1 (\mathbf{y}^* - \mathbf{y}) + \mathbf{k}_2 \int_{t_0}^{t} (\mathbf{y}^* - \mathbf{y}) dt$$

$$(2.26)$$

where y* represents the process setpoints, and the positive values of λ^-_p or λ^+_p represent the difference between the output response and the GMC reference trajectory. The constrained multivariable control problem can be solved as a non-linear constrained optimization problem, which minimizes a function of the slack variables.

Lee and Sullivan (1988) provided a more detailed explanation and study of the properties of this control law. It is important to note that the control law expressed in Equation 2.22 directly imbeds an approximate dynamic, possibly non-linear process model as defined in Equations 2.19 and 2.20. Any mismatch between the predicted plant behaviour from the process model and the real process will be compensated for by the integral term in the control law. This term ensures not only robust process behaviour in the presence of modelling error, but also the elimination of steady-state offset. A more detailed review is given in Chapter 6 on constraint handling capabilities of GMC with real-time application to catalytic reforming process. In Chapter 7, a comprehensive review was given in relation to a more difficult process of fluid catalytic cracking, with multiple constraints and process interaction.

2.6.2 GMC and Internal Model Control (IMC)

Relationships between GMC and IMC are discussed in this section. It is shown by Lee (1991) that a special form of filter in IMC can lead to GMC if the process model is perfect. However, this special filter results in the direct application of not only linear but also non-linear process models.

The internal model control structure and a filter in the z-transform domain was shown in Figure 2.3. Under the assumptions that the model is perfect, the process **G** can be factorized as:

$$\mathbf{G} = \mathbf{G_+} \mathbf{G_-}, \qquad \mathbf{G}(0) = \mathbf{I} \tag{2.27}$$

where $\mathbf{G_+}$ contains all process deadtimes and RHP zeros such that $\mathbf{G_-}$ has a stable and realizable inverse. The controller is chosen as:

$$\mathbf{G_C} = \mathbf{G}_-^{-1} \tag{2.28}$$

The closed-loop response for the reduced IMC structure is given by:

$$\mathbf{u} = \mathbf{G}_-^{-1} (\mathbf{y^*} - \mathbf{d}) \tag{2.29}$$

$$\mathbf{y} = \mathbf{G_+} (\mathbf{y^*} - \mathbf{d}) + \mathbf{d} \tag{2.30}$$

The block diagram of the reduced IMC structure is shown in Figure 2.5. In this figure $\mathbf{d}(s)$ is denoted as disturbances from the process and $\mathbf{d'}(s)$ is denoted as errors in the model.

The GMC control law is:

$$\dot{\mathbf{y}}_M = \mathbf{k_1} (\mathbf{y^*} - \mathbf{y}) + \mathbf{k_2} \int (\mathbf{y^*} - \mathbf{y}) dt \tag{2.31}$$

where the subscript M indicates that the left hand side is calculated by the model. For linear system, taking Laplace transform on Equation 2.31.

$$\dot{y}_M(s) = (k_1 + \frac{1}{s} k_2) (y^*(s) - y(s)) \tag{2.32}$$

If the model is perfect *i.e.*, there is no process-model mismatch, the following equations are true:

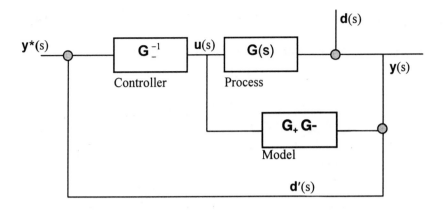

Figure 2.5. The block diagram of the reduced IMC structure

$$y(s) \quad = G(s)\, u(s) + d(s) \tag{2.33}$$

$$\dot{y}_M(s) = sG(s)\, u(s) + Ed(s) \tag{2.34}$$

where **E** is a constant matrix with appropriate dimension. Substituting Equations 2.33 and 2.34 into Equation 2.32 yields:

$$\mathbf{G}\, u = (s^2 I + s k_1 + k_2)^{-1} [(s k_1 + k_2)(y^* - d) - s E d] \tag{2.35}$$

In view of making the controller stable and realizable, the same factorization as Equation 2.27 is adopted and the GMC controller becomes:

$$u = \mathbf{G}_-^{-1}\, (s^2 I + s k_1 + k_2)^{-1} [(s k_1 + k_2)(y^* - d) - s E d] \tag{2.36}$$

and the closed loop response is:

$$y = \mathbf{G}_+\, (s^2 I + s k_1 + k_2)^{-1} [(s k_1 + k_2)(y^* - d) - s E d] + d \tag{2.37}$$

Define F_1 and F_2 as:

$$F_1 = (s^2 I + s k_1 + k_2)^{-1} (s k_1 + k_2) \tag{2.38}$$

$$F_2 = (s^2 I + s k_1 + k_2)^{-1} (sE) \tag{2.39}$$

then the controller and the closed-loop response can be rewritten as:

$$u = \mathbf{G}_-^{-1}\, [F_1 (y^* - d) - F_2\, d] \tag{2.40}$$

$$y = G_+ [F_1 (y^* - d) - F_2 d] + d \qquad (2.41)$$

The block diagram of this reduced version of GMC is shown in Figure 2.6.

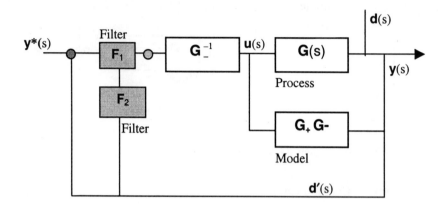

Figure 2.6. The block diagram of the reduced GMC structure

It is evident that the reduced version of GMC is equivalent to the reduced IMC plus the two second order filters F_1 and F_2. However, these two filters enable the direct application of both linear and non-linear state space process models within a GMC controller. In Figure 2.6, $d(s)$ are the disturbances from the process and $d'(s)$ are the errors from the model. It is also clear from Equation 2.37 that the closed loop response is stable if the process G is stable. The condition on the factorization of G has ensured the stability of the controller in Equation 2.36. It is noted that for linear systems where the model is not perfect there is no simple method to rearrange the GMC structure into the IMC structure.

2.6.3 GMC and Model-predictive Control (MPC)

2.6.3.1 Discrete form of GMC

Model-predictive control, as reviewed in Section 2.2, is a discrete control method based on the convolution model. The future process outputs are predicted by the past, present and future control actions through the convolution model. Then the present and future control actions are determined by minimizing the difference between the predicted outputs and the reference trajectory, which is defined by the present output and the setpoint. Model algorithm control (MAC) (Mehra *et al.*, 1979) and dynamic matrix control (DMC) (Cutler and Ramaker, 1980) are two of the special cases of MPC. In the following, the discrete form of the GMC control law is first derived in order to compare GMC with MPC.

Suppose a linear system is discretized as:

$$x_{k+1} = \Phi x_k + \Gamma u_k \qquad (2.42a)$$

$$y_{k+1} = C\, x_{k+1} \tag{2.42b}$$

where the subscript k denotes the k-th sampling time.

The GMC control law

$$\dot{y} = k_1\,(y^* - y) + k_2 \int (y^* - y)dt \tag{2.43}$$

can be discretized as:

$$\frac{y_{k+1} - y_k}{\Delta t} = k_1\,(y_k^* - y_k) + k_2 \sum_{i=1}^{k} (y_i^* - y_i)\Delta t \tag{2.44}$$

where Δt is the sampling time. Equating Equations 2.42 and 2.43 and assuming that $C\Gamma$ is invertible, yields the following:

$$u_k = (C\Gamma)^{-1}\,[\, y_k + \Delta t\,(k_1\,(y_k^* - y_k) + k_2 \sum_{i=1}^{k} (y_i^* - y_i)\Delta t)$$
$$ - C\,\Phi\, x_k\,] \tag{2.45}$$

Equation 2.45 is one of the discrete forms of the GMC law for systems described by Equation 2.42. The velocity form of Equation 2.45 is:

$$\Delta u_k = (C\Gamma)^{-1}\,[\, y_k - y_{k-1} + \Delta t\,(k_1\,(y_k^* - y_{k-1}^* - y_k + y_{k-1})$$
$$+ k_2\,(y_k^* - y_k)\,\Delta t) - C\,\Phi\,(x_k - x_{k-1})] \tag{2.46}$$

for constant setpoint ($y_k^* = y^*$), the above equation reduces to:

$$\Delta u_k = (C\Gamma)^{-1}\,[\, y_k - y_{k-1} + \Delta t\,(k_1\,(- y_k + y_{k-1})$$
$$+ k_2\,(y_k^* - y_k)\,\Delta t) - C\,\Phi\,(x_k - x_{k-1})] \tag{2.47}$$

2.6.3.2 Relationship between Discrete GMC and MPC
The MPC control law (Marchetti *et al.*, 1983) for MIMO system is given by:

$$\Delta u = (A^T Q^T Q A + r^2 I)^{-1} A^T Q^T Q\, e \tag{2.48}$$

where **A** is the "dynamic matrix" which is derived from the impulse response convolution model of the system. **Q** is the weighting matrix on the predicted output error over the prediction horizon. r^2 is the weighting factor which penalizes manipulated variable moves (Δu). Only Δu_k from Equation 2.48 is implemented at each sampling time. Therefore,

$$\Delta u_k = K \, e \qquad\qquad (2.49)$$

where the moves of manipulated variables are implemented at each sampling point (Δu_k) and k is a constant matrix, which is the appropriate part of the dynamic matrix $(A^T Q^T Q A + r^2 I)^{-1} A^T Q^T Q$.

Comparing Equation 2.49 with Equation 2.47, the following conclusions can be drawn:

1. The GMC control law is based on the present and historic output measurements. The control movement (Δu_k) based on these measurements and the process model provide the necessary control action to achieve the pre-defined GMC reference trajectories of the outputs.
2. The MPC control law is based on the present output measurements and the predicted future outputs. The predictions are made through the convolution model using the historic control action. The control movement provides the necessary control action to achieve the DMC reference trajectories of the outputs.
3. The GMC reference trajectories are defined based on the initial system outputs and the setpoints. Therefore, the present control aim is always associated with the past system outputs (through the integral term). On the other hand, the MPC reference trajectories are defined based on the present system outputs and the setpoints at each sampling time.

2.7 Stability and Robustness of Non-linear Control

Stability and robustness of a controller under process uncertainties is recognized as one of the most important issues in process control. In control theory, it is well understood that the closed-loop system must be stable, which ensures that the system variables remain bounded. The closed-loop system is studied based on a proposed process model, and the stability can only be guaranteed if the process is exactly described by the model. For process control, the concept of robust stability, defined as "the stability of the closed-loop system in the face of the uncertainties in the process model (process/model mismatch)", is more applicable as it is rarely true that the model can represent the process exactly.

The robustness analysis of feedback systems has attracted considerable interest for both linear systems (Safonov, 1980; Doyle and Stein, 1981; Doyle, 1982; Morari and Zafiriou, 1989) and non-linear systems (Kravaris and Palanki, 1988; Calvet and Arkun, 1989; Nikolaou and Manousiouthakis, 1989; Khambanonda *et al.*, 1990). Important theoretical properties such as closed-loop stability and performance are still under investigations even for linear systems. For example, Rawlings and Muske (1993) recently showed the conditions under which linear model-predictive control (LMPC) guarantees closed-loop stability in the presence of feasible constraints. In the case of non-linear systems, as noted by Beigler and Rawlings (1991), many important questions such as nominal stability, robust stability, and performance are unanswered.

Chen and Shaw (1982) used the final state constraint to show that a continuous-time non-linear system is stabilized by a receding-horizon control (RHC) formulation. Mayne and Michalska (1990) provided a rigorous mathematical treatment of stability of RHC for continuous-time non-linear systems.

Henson and Seborg (1993) assumed that non-linear transformation analogous to results in continuous-time differential geometry-based approaches are available to analyze the closed-loop properties. The only other approach to study the closed-loop properties of non-linear model-based control algorithms was provided by de Oliveira and Beigler (1995).

Meadows *et al.* (1995) studied the nominal closed-loop stability issues using a Lyapunov function type of analysis. de Oliveira and Beigler (1995) extended the Newton-type control algorithms of Li and Beigler (1989) to a predictive control framework. The main stability result in their work proves that the nominal closed-stability can be achieved for open loop stable system provided a sufficiently long sample time is chosen. Their work, however, assumes that all states are measured and does not provide guidelines for selecting the required sampling time.

Sistu and Bequette (1996) focussed on a methodology for closed-loop stability analysis that does not require any explicit terminal state constraints used in the RHC formulation in order to guarantee closed-loop stability. Their work on stability analysis is based on computable criteria for a system composed of a non-linear process and a discrete controller. The state and input sensitivity equations of the continuous-time model are used in computing the nominal closed-loop stability criteria. A simulation example of a CSTR with input multiplicity was presented illustrating the analysis methods and closed-loop behaviour.

Zhou and Lee (1995) studied the problem of stability analysis and robustness for non-linear control system within the framework of generic model control (GMC). It was demonstrated that GMC is robust for limited process/model mismatch. Stability and robustness of GMC and its extensions were analyzed under the condition that explicit control law is available. For non-linear control systems, robust stability synthesis of GMC was studied by input-output theory. Based on the passivity theorem, a procedure was proposed to synthesize a stable GMC controller for a family of processes. Processes need to be properly scaled before using this procedure. The controller parameters must be selected so that a modified process and a modified controller satisfy the conditions of the passivity theorem.

The same authors also showed that by incorporating an open-loop observer in GMC would implicitly incorporate integral control action. Therefore an explicit integral control term in the GMC control law can be eliminated. With such elimination, the closed-loop system fits into the structure of classical feedback control and its analysis is simplified. The detailed discussion and derivation of stability conditions can be found in Zhou and Lee (1995).

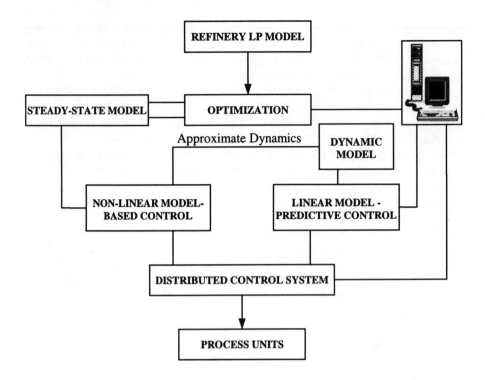

Figure 2.7. Overview of non-linear model-based control system hierarchy

2.8 Conclusions and Discussion

In this review chapter, the current interests of the processing industry and the latest trends in academic research in model-based control have been highlighted and many industrial applications of model-predictive control have been reported. The need for non-linear model-based control is emphasized, as model-based control algorithms based on linear model can not handle non-linear systems very well, especially when the non-linearity becomes strong. Therefore, a model-based control algorithm based on non-linear model offers more promise in practice for the majority of refinery process control problems. Stability and robustness issues related to both the linear and non-linear control systems are also discussed. A brief review of GMC system (linear and non-linear) is given and compared with internal model control (IMC) and model-predictive control.

In this chapter, relations between generic model control (GMC) and internal model control (IMC) and model-predictive control (MPC) have been investigated. It was found that for linear systems, if the model is perfect, GMC is equivalent to IMC plus two special filters. These two filters lead to the direct application of state space process model rather than transfer functions. *This also implies that GMC is applicable to both linear and non-linear systems.* The discrete GMC control law was derived for a discrete process model. It was concluded that different schemes of differential and integral approximation would result in different forms of the control law. The major difference between GMC and MPC is in the usage of past process measurements and past control action, which is partly due to the different output reference trajectories and different forms of the process model.

In this chapter, it has been shown that the majority of model-based control systems are based on a linear model of the process. The GMC and non-linear IMC (NL-IMC) can directly employ a non-linear model within their control law. However, in non-linear IMC (NL-IMC) it is required to calculate the inverse of an operator containing memory. This operator is the process model; a set of differential and algebraic equations. In order to calculate the GMC control action using the explicit formulation, it only requires the solution of a set of non-linear algebraic equations. That is, if GMC is regarded as a problem to calculate the inverse of an operator, then this operator is a static operator without memory. Therefore, implementation of GMC is much easier than that of NL-IMC. Moreover, the parameters of GMC have an explicit and easily understood effect on the closed-loop responses.

Finally, the non-linear GMC is selected for further research, design and development work in this text and for the non-linear control applications in real-time to a wide range of refinery processes. Figure 2.7 gives an overview of non-linear model-based control system hierarchy. The system shows that the linear model-predictive control requires dynamic model from the process identification tests and the steady-state model for the use of optimizer. In non-linear model-based control, the controller and optimizer share the same model. This provides an improved control performance by minimizing the process-model mismatch and reducing the plant testing time for process dynamics identification.

CHAPTER 3

INFERENTIAL MODELS IN NON-LINEAR MULTIVARIABLE CONTROL APPLICATIONS

On-line product quality measurements suffer from excessive time delays and sampling intervals, which limit their effectiveness as inputs for automatic control. Inferential models which are based on fast and continuously available temperature, pressure and flow measurements reduce the negative impact of the sample intervals and time delays with minimum compromise on accuracy. The resulting continuous and fast response keeps the product qualities on specifications and minimizes the quality giveaway. This chapter provides the development procedure of correlation-based inferential models for a wide variety of refinery processes, use of inferential models in both open and closed loop environments, laboratory update mechanism, operators acceptance and benefits of inferential models and their use as a controlled variable in non-linear multivariable control applications.

3.1 Introduction

An inferential model can be seen as a simple process model, which correlates one or more off-line measured product compositions/properties with on-line measured data such as temperatures, pressures and flows. It generates continuous real-time process stream property data, which are calibrated regularly via manual insertion of off-line quality/property measurements. Off-line and, to a lesser degree, on-line product quality measurements suffer from excessive time delays and sampling intervals which limit their effectiveness as inputs for automatic control. Inferential models, based on fast and continuously available temperature, pressure and flow measurements, are developed to reduce the negative impact of these sample intervals and time delays with minimum compromise on accuracy. The resulting continuous and fast response helps operators and controllers to monitor the actual product qualities.

The pioneering work in inferential control was presented by Joseph and Brosilow (1978) and Brosilow and Tong (1978). Later work improved much of the statistical aspects, specifically the work in the area of partial least squares (PLS), principal component analysis (PCA) and ridge regression. The Brosilow estimator was the first explicit reference to inferential control though earlier applications of various types had been reported by Brosilow and Tong (1978). The multivariate statistical methods such as PLS provide an obvious basis for a general solution to building inferential models and these methods have been used effectively by a number of authors. Wright *et al.* (1977), and Mejdell and Skogestad (1990) used principal component regression (PCR) to develop inferential control models for the multivariable adaptive control of a packed bed reactor, and for the control of a linear binary distillation column. Moore *et al.* (1986) used PCA to determine the selection of the best tray temperature for distillation column control, while Kresta *et al.* (1990) used PLS for building inferential models for distillation column control which retained most of the measured variables.

Parrish and Brosilow (1988) developed non-linear inferential control (NLIC) for improving control of non-linear systems. The controller developed was model based, and allowed for direct use of available measurements. The major contribution of their work was to use available measurements to infer how disturbances have influenced the process. It also provided a decomposition of the control problem, which facilitated the design of the non-linear control system and provided a conceptually simple and practical method for tuning the control system to accommodate the inevitable modelling errors.

The major conclusions from these studies were that the multivariate statistical estimators provided excellent performance in predicting the product qualities. By using many measurements in the model much better predictions were obtained, because a wider range of unmeasured disturbances could be detected, and because of the noise averaging effects. Furthermore since PCA and PLS easily handle missing data, the inferential models based on them are much less sensitive to sensor failures. Mejdell and Skogestad (1990) showed in their distillation simulations that a steady-state PCR inferential model performed as well as the dynamic Kalman Filter model. Kresta *et al.* (1990) emphasized the importance of data collection method to develop the model. It is important that the data set contain information not only on the effect of the manipulated variables, but also on the important disturbances in the process. Even more important is the necessity of collecting the data under closed loop conditions that resemble as closely as possible the conditions under which the model will be used. Another common industrial situation is that the information for the plant is available at several sampling rates. In distillation column control, for example, one might like to use tray temperature measurements combined with occasional measurements from a product analyzer, such as a gas chromatograph. Lee *et al.* (1990) proposed a multi-rate model-predictive control (MPC) formulation for such problems.

Baratti *et al.* (1995) reported the development of non-linear extended Kalman filter, which infers the compositions of the streams leaving a binary distillation column from temperature measurements. The accuracy of the distillation column model, on which the estimator was based, was discussed in connection with the reliability of the obtained estimates. The estimator performance was checked by comparison with the dynamic behaviour of a distillation pilot plant.

Recently Dayal and MacGregor (1997) have developed a new and fast recursive, exponentially weighted partial least squares algorithm which provided greatly improved performance for the industrial soft sensor application as compared to the recursive least squares algorithm.

Neural networks are another interesting example of developing the inferential model for chemical process. Software based on the neural network approach called "Process Insight" by Pavilion Technologies, Inc. was developed by Ungar *et al.* (1995) and is available commercially. The main features of this neural network-modelling tool are that it has the ability to build large models and because of the structure of the neural networks, it captures non-linear relationships between variables.

As discussed above, there are a number of methods available to infer product quality data on-line. These include correlations based on laboratory data, pressure compensated temperature, heat and material balance calculations, and correlations based on steady-state models. The methods differ in their complexity and ability to predict product qualities accurately. The traditional "material and energy balance" method of calculating properties, which is most commonly used by operating companies, does not perform well for the middle distillate range material on units such as atmospheric and vacuum distillation, FCCU and hydrocracker main fractionators. One problem with this approach is that many of the trays never reach phase or thermal equilibrium because of super-heated feed vapours. The technology developed by DMCC (1993) enhanced the performance of the distillation models to enable non-equilibrium tray-by-tray calculations and has provided improved results. The above method works well to predict the desired product quality. However, without updating the on-line laboratory bias correction, the on-line calculations may not adequately match the laboratory values.

The main contribution of this work is to develop a wide range of inferential models using correlation based techniques. The steady state models of the fractionator and debutanizer were used to develop reliable inferential calculations for product qualities. A number of case-studies were made here with these models and then regressed to develop the relationships between operating data and product qualities. These relationships are implemented on-line and updated with laboratory data to ensure agreement between the calculated values and laboratory data.

The inferential models predicting the RVP and iso-pentane of product streams were incorporated into non-linear multivariable control strategies of an industrial debutanizer in Chapter 4. These models formed a closed-loop quality control system reducing the product giveaway. The octane inferential model developed here was incorporated into constrained non-linear multivariable control strategies of a catalytic reforming process in Chapter 6. This model formed a closed-loop WABT/octane control system and when implemented in real-time, it reduced the octane giveaway. In addition, the use of inferential models eliminated the requirement of expensive analyzers, cost and efforts to maintain these analyzers.

3.2   Development of Inferential Model

3.2.1   Overview of Models

The development of a good inferential model begins with a thorough understanding of the process. This allows the essential cause and effect variables to be identified.

A general theme in inferential control is that in the absence of analyzers much is expected from an inferential model. It should be able to detect feed composition changes, disturbances, and be able to estimate outlet compositions under normal operations. Process experience shows that a large number of process inputs need to be considered before the behaviour of the unit can be predicted. The preferred process measurement is temperature, and therefore many forms of inferential controllers are built around multiple temperature measurements.

To develop an inferential model for cutpoint control, for example, we consider the following explicit form of equations:

$$\hat{y}(s) = K(s)\, \tilde{Q}(s) \qquad\qquad (3.1)$$

$$\tilde{Q}(s) = [y, t, u, d] \qquad\qquad (3.2)$$

where $\hat{y}(s)$ is the dependent variable, overhead composition in a distillation column and $K(s)$ is a correlation coefficient and $\tilde{Q}(s)$ is a vector which contains feed rate y, tray temperature t, draw rates u and composition disturbances d. At this stage, the variable d could be set to zero if it is known, as the effect of feed composition changes is difficult to model, and therefore the disturbances can be quickly observed in the change of temperature profile. Skogestad *et al.* (1990) have shown the adverse effects of including the vectors **u** and **d** in Equation 3.2, using inferential models applicable to linear systems. However, the inferential models developed in this paper can be used in both linear and non-linear applications for a wide range of operating conditions in order to control the product qualities.

There are two kinds of model that can be developed. The first type is called a parametric model, and the other is based on correlation. The parametric model is generally derived from first principles, and the coefficients are included in various functional forms to generalize the model. The major groups in the models are extracted from first principles, with enough degrees of freedom remaining to allow it to be quickly applied to similar unit operations. This type of model has been used successfully with binary columns, but it is difficult to generalize for multicomponent columns. The main reason is that too many dimensionless groups have to be introduced, making this model non-robust. Such a parametric model was described by Jafarey *et al.* (1980) in the following form:

$$V/F = \frac{\Phi\, R_{min}\ \ D/F}{1 + \Theta} + \Psi \qquad\qquad (3.3)$$

where V/F is the ratio of vapour to liquid flow, D is the distillate flow, R_{min} is the minimum reflux to the column, Φ is a fudge factor or degrees of freedom and ψ is a bias (or a trim factor) used to update the model.

3.2.2   Non-linear Inferential Control Model

The non-linear inferential control (NLIC) model developed by Parrish and Brosilow (1988) relates the control effort and disturbances to the process measurements. The models considered by these authors may be expressed as:

$$X_k = f(X_{k-1}, m_{k-d}, U_k, A) \tag{3.4}$$

$$Y_k = h(X_k) \tag{3.5}$$

$$\theta_k = \theta(X_k, m_{k-d}, U_k) \tag{3.6}$$

where X is the state vector of dimension n, y is the model output, θ is the process measurements of dimension r, U is the input disturbances of dimension p, A is the parameter vector of dimension q, n is the control effort, d is the control effort time delay (d>0) and k is the time index.

The following conditions are placed on the process model in order that it can be used with the NLIC. These conditions arise so that NLIC reduces to LIC when a linear model is used, and that the stability of the control system can be assured for large enough values of the filter tuning parameter.

Condition 1: the partial derivatives with respect to X, m, U, A must all exist and be finite throughout the operating region (X, m, U, A). This condition allows the model, given by Equations 3.4 and 3.5 to be linearized about all operating points.

Condition 2: the process model must be locally asymptotically stable everywhere in the operating range.

Condition 3: the sign of the steady state gain, k, must not change in the operating region. That is $|k| > 0$. Condition 3 is a necessary condition for the existence of a stable control system with no steady state offset of the controlled variable from its setpoint (Parrish, 1985).

Condition 4: the model must be capable of exhibiting the measured process response in spite of process modelling errors. This property is referred to as "model consistency" and is required so that the model response can follow the process.

3.2.3   Correlation-based Model

The inferential models developed here are based on correlations derived from the steady-state models for an actual process. The expression may be linear or non-linear, but it was kept as simple as possible to enhance robustness and maintainability. When the models are non-linear, the regression can still be carried out from process data, because the equations are still linear with respect to the coefficients. For a crude distillation application, the reference temperature was at some distance from the tower ends. The absolute values of temperatures were used due to large temperature differences in the top section of the column due to process disturbances.

On the crude column, the naphtha final boiling point (NFBP) prediction is based on the following correlation derived from the steady-state model of crude distillation, and retaining the process knowledge in the controller:

$$NFBP = [\ P, T, \Lambda_1, R/D \ ] + \Lambda_2 \qquad\qquad (3.7)$$

In Equation 3.7, P and T are the column pressure and top tray temperature respectively, R/D is the reflux to distillate ratio, Λ_1 is the filter of the first order to reduce the measurement noise and Λ_2 is a bias or trim factor to update the model from the laboratory quality results. It is important to note that Equation 3.7 is non-linear in terms of the reflux:distillate ratio. The advantage of using a non-linear model is that it is valid over a wider operating range and reduces the retuning effort. Equation 3.7 above does not take into account pressure-temperature compensation. The temperatures in distillation columns are normally pressure compensated, therefore, the pressure-temperature compensation was carried out using the non-linear Clausius-Clapeyron equation given in Perry and Green (1984). It was modified in the following form for this particular application:

$$Tcomp = T_k * B_i \ /(T_k * LnP + B_i \ ) \qquad\qquad (3.8)$$

where Tcomp is a pressure compensated temperature, T_k is tray temperature in Kelvin and B_i is a Clausius-Clapeyron coefficient. Equation 3.7 was derived from the steady-state model, which receives the scanned values of pressure, temperature, reflux and distillate every two minutes from the real-time data acquisition system. Based on these values the model calculates and predicts the quality of the product. The dynamic effects are compensated as the tray temperature at the top of the column and reflux and distillate flows are fast loops. The effect of pressure variation in the column is incorporated by the pressure-compensated temperature obtained from Equation 3.8. The application programme to obtain the pressure compensated temperature from Equation 3.8 was written in control language (CL/AM) and implemented in Honeywell's TDC-3000 DCS application module (AM). Equation 3.8 was modified to work in real-time application and the programme is given in Appendix A. The calculations to obtain the Clausius-Clapeyron coefficient (also denoted as CC in Equation A-1, implemented in real-time) is given in Appendix A.

3.2.4   Verification of Correlation-based Model

It is important to verify the correlation-based model, and the determination of its parameters against actual process behaviour. The expression may not provide good predictions if the data collected under normal operating conditions does not contain sufficient parameters. Numerical solutions to solve this type of expression will introduce sufficient error, that even the signs of the model coefficients may be incorrect. There are two ways to overcome this problem. It has been shown by Parrish (1985) that the coefficients can be identified off-line, and incrementally updated on-line using Kalman filters. Malik (1991) suggests that the coefficients of the estimator are determined off-line by a steady-state model and kept fixed during on-line operation, then a bias parameter can be used to obtain steady-state agreement with the process.

3.3   Online Applications of Inferential Models

3.3.1   Naphtha FBP of Crude Distillation

The application programme to solve the FBP inferential model was written in Control Language (CL/AM) and implemented in AM (Application Module) of Honeywell TDC-3000 DCS. The details of the programme and calculations are given in Appendix A. The inferential model developed predicts the Naphtha FBP in crude distillation. Figure 3.1 shows the systematic approach used to construct the Naphtha FBP inferential model with process variables from crude distiller such as a pressure-compensated temperature, pressure of the column and reflux ratios.

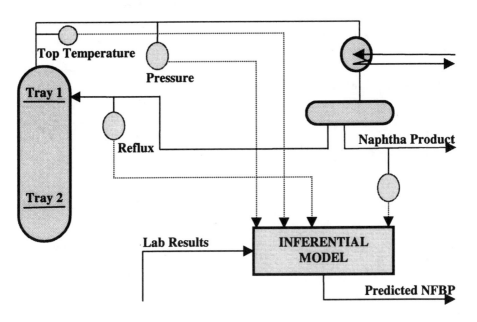

Figure 3.1. Inferential model development procedure for the NFBP system

The model is updated using laboratory results. A system of inferential model update is shown in Figure 3.2, providing laboratory result as a basis to update the predicted values and a first-order filter to compensate for the noise in the flow measurements. The average deviation from the laboratory results is only 2 °C, and considering the repeatability of ± 4 °C in naphtha FBP laboratory results, this prediction can be considered very accurate. Figure 3.3 gives a comparison between the laboratory results and the prediction from the model for naphtha FBP. It must be noted that the evaluation of the prediction statistics was done only at the time when the laboratory sample was inserted in the system.

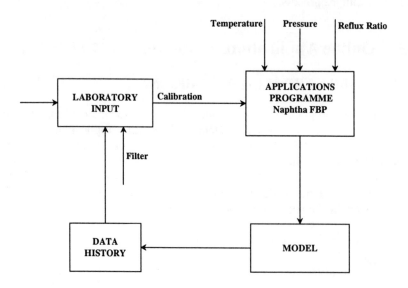

Figure 3.2. System of inferential model update

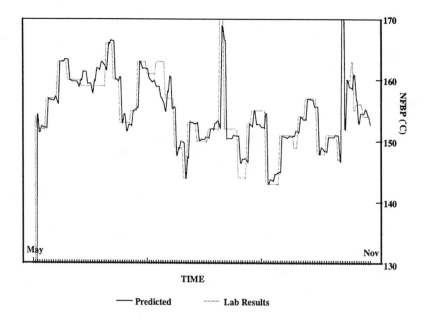

Figure 3.3. Comparison of NFBP prediction with the laboratory results

The naphtha-initial-boiling-point (NIBP) inferential model follows the same method as discussed in the development of naphtha FBP previously described. The difference in developing a suitable correlation can be found in the use of reboiler ratio instead of reflux ratio.

After implementing the naphtha FBP inferential model in real-time, it was observed that the model prediction was deviating significantly from the measured value of the product quality (laboratory results) under varying process conditions (process-model mismatch, changes in the naphtha mode of operation).

In order to overcome this difficulty, the model was modified (revised) to take the form of a module which updates the bias term Λ_2 in Equation 3.7 of the prediction model in response to changing condition and model inaccuracies. This module is called the "variable bias module" which enhanced the predictive ability of the existing inferential model.

Figure 3.4 shows the improvement in the results of revised inferential model with variable bias module. The revised model now has the ability to predict accurately in the face of varying operating conditions. In addition, the inferential model also predicted accurately in the absence of laboratory results so long as naphtha mode of operation was not changed.

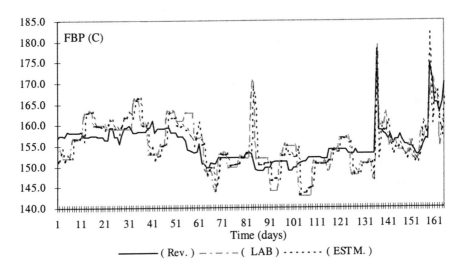

Figure 3.4. Results of inferential model prediction (Rev.) with variable bias module

3.3.2 Kerosene Flash Point of Crude Distillation

The kerosene flash point inferential model was developed based on correlation between kerosene stripper reboil ratio and pressure compensated tray temperature (T_{comp}). This was an interesting application as the distillation column operated in three different modes of operation: keroflash (40°C); keroflash (50°C) and keroflash (65°C). Therefore it was necessary to select the variables that cover the entire range of operation. The best choice proved to be a combination of two important variables:

- The ratio of reboiler duty to side product (kerosene flow); the duty itself was found to depend too much on the mode of operation.
- Pressure compensated temperature of tray 14 in the column.

Initially it was expected that the top reflux might affect the kerosene flash point but process identification tests on this column revealed that the kerosene flash point was in fact independent of the top reflux and therefore this variable was removed from the correlation expression. The functional form of the correlation of kerosene flash point is given as:

$$KeroFlash_{model} = f(C_C, T_{comp}, C_s, Reboil\ duty, KeroFlow) + C_3 \qquad (3.9)$$

C_C = a parameter weighting the pressure compensated temperature which is a measure for the heaviness of the stream.
C_S = a parameter weighting the reboil ratio which is a measure for the separation sharpness in the column.
C_3 = a calibration constant, trimmed by laboratory results.
Figure 3.5 shows the on-line prediction of the kerosene flash point updated once a day by laboratory results.

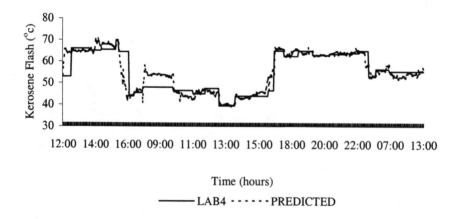

Figure 3.5. Kerosene Flash Point inferential model predicting the flash point of kerosene at three modes of operation and compared with laboratory results

It may be observed from Figure 3.5 that the inferential model covers the full range of operating modes of the kerosene flash point. The performance of the model was excellent. The difference between laboratory and predicted results was within 2°C in most of the operating modes.

3.3.3   RVP of Debutanizer bottom

The correlation developed for RVP of platformer feed of a debutanizer is a simple correlation based on pressure (P), temperature (T) and 10% ASTM distillation. The laboratory results (C_1) were used to update or trim the model. The results were compared with laboratory results as shown in Figure 3.6. A normalized distribution was assumed and standard deviations were calculated, based on the observed "laboratory minus predicted" difference. For the most stable plant data, RVP prediction exhibited a standard deviation of 0.01 bar. This inferential model was first used as an off-line model and then incorporated in non-linear model-based multivariable control system, forming a closed-loop quality control system in real-time application to an industrial debutanizer in Chapter 4. The functional form of the correlation is given below:

$$RVP_{model} = f\,(10\%\ ASTM\ distt., P, T,) + C_1 \qquad\qquad (3.10)$$

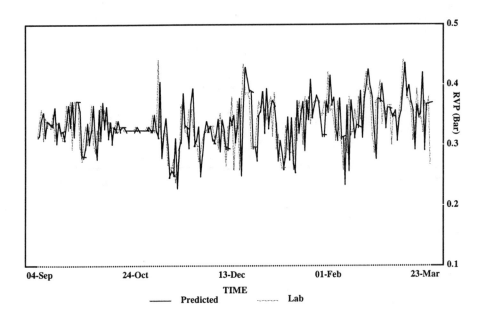

Figure 3.6. Comparison of RVP of platformer feed prediction

3.3.4   Iso-Pentane of Debutanizer Overhead

The correlation developed for iso-pentane from a debutanizer overhead is based on pressure-compensated temperature, non-ideal gas correlation factor and a bias. The analyzer results were used to update or trim the model. In the case of iC_5, analyzer results were compared with the inferential model predictions and the results are very accurate as shown in Figure 3.7. It is important to mention that the changes predicted by the inferential model were sometimes of the same magnitude as the recognized repeatability of the laboratory test. On-line analyzer results, when available, generally agreed with the predicted values, but they exhibited long time delays.

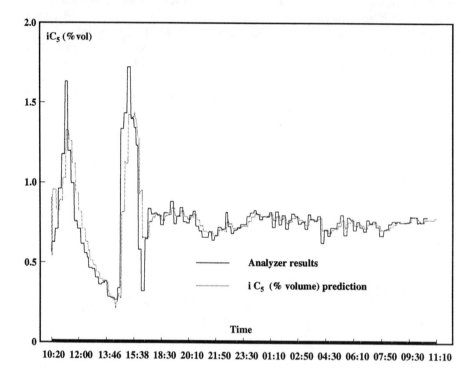

Figure 3.7. Comparison of debutanizer overhead iso-pentane prediction with analyzer results

The analyzer measurements were not integrated into the model due to these long time delays (15-20 minutes). Instead we have used analyzer results or laboratory analysis to update the model if its standard deviation exceeded the specified limits. The correlation was developed from the real-time data of a debutanizer, collected over a period of time under varying conditions. Since the correlation was derived from the steady state model of debutanizer, inferential calculations were also based on the properties of typical debutanizer, and laboratory results were used to tailor the inferential model results to the actual debutanizer operation. The final form of correlation derived is given in Appendix A and the steady state model of debutanizer is discussed in Chapter 4.

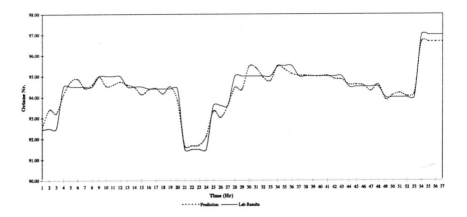

Figure 3.8. Comparison of reformate octane inferential model prediction with laboratory results

3.3.5 Octane Inferential Model for Catalytic Reforming

The octane inferential model development is the most difficult and challenging task due to the fact that a lot of variables such as chloride levels, space velocity, feed compositions, operating severity, hydrogen partial pressure, feed compositions, recycle hydrogen rate, reactor temperature, reactor system pressure and carbon on catalyst *etc.*, can influence research octane number (RON) of reformate. Therefore the model was kept as simple as possible and only those variables were considered in the correlation model which were necessary and contributed to octane number. Finally the model takes the following form of correlation:

$$\text{Reformate octane} = f\,(PONA, P, T, F, RG, C_1) + C_2 \qquad (3.11)$$

In Correlation 3.11, PONA is the feed composition of the reformer feed (paraffins, naphthenes, olefins and aromatics), P is the system pressure, T is the reactor temperature, F is the feed rate, RG is the recycle gas rate, C_1 is the first-order filter and C_2 is the laboratory bias. The model was used in a feedforward mode to adjust the target WAIT based on changes in the disturbance variables.

The laboratory results were used to update the model based on the difference between the model prediction at the time the sample was taken and that the actual measured value. Figure 3.8 shows the inferential model predicting reformate octane. The average deviation from the laboratory results is only ± 0.2 octane number, considering the repeatability of ± 3 octane number in laboratory results, this prediction can be considered very accurate. This octane inferential model was used in real-time and formed a closed–loop quality control system with the constrained non-linear multivariable control application to the catalytic reforming process in Chapter 6.

3.4   Tuning of Inferential Models

In order to tune the inferential models to the plant operation, it was necessary to collect several sets of plant data and corresponding laboratory results. Unsatisfactory mass balance errors or unstable unit operation led to the rejection of some of the data collected. In addition, data were accepted only if enough laboratory results were available to completely specify the products in terms of their distillation curves. When these conditions were satisfied, the following tests were performed.

- *Repeatability Test* This test is an on-line sensitivity analysis. The changes in model predictions, on a run-to-run basis, were compared quantitatively with plant input data variations. Tests were conducted for "steady state" unit operation to ensure that normal variations did not lead to unacceptable variations in model predictions.
- *Directional Response Test* This test was conducted to observe the model responses during changes in yield, crude feed and fractionation.
- *Stability Test* This test was carried out concurrently with the directional response test. All model predictions were compared with plant and laboratory data over a long time period (4 weeks).
- *Accuracy Test* The accuracy of the model predictions was tested against plant and laboratory data. Assessment of the results considered ASTM test accuracy and repeatability derived from studies undertaken regularly in the laboratory.

Table 3.1 Comparison of standard deviation of the inferential model prediction and laboratory results; and the manual control and update only.

Variables to Control	Max. Allowable Standard Deviation	Standard Deviation Model Pred./Lab. Results	Standard Deviation Manual Control / Lab. Update
Iso-Pentane	0.1 % vol.	0.06 % vol.	0.2 % vol.
Naphtha Final Boilng Point	3 °C	1.5 °C	4.0 °C
Naphtha Initial Boiling Point	3 °C	1.5 °C	4.0 °C
Kerosene Flash Point	3 °C	2 °C	4 °C
Octane Number	± 3	± 2	–
Reid Vapour Pressure	0.02 bar	0.01 bar	0.03 bar

Table 3.1 summarizes the maximum allowable standard deviation of the difference between the inferential model prediction and the laboratory analysis for any given quality over all valid sets of results. This table also shows the actual standard deviation between the inferential model prediction and the laboratory

results and between the manual control and laboratory results for all the product
qualities discussed here.

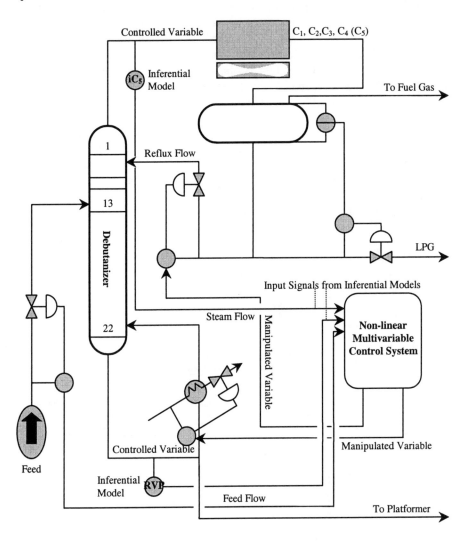

Figure 3.9. Use of inferential models in non-linear multivariable control of the crude
debutanizer

3.5 Inferential Models in Non-linear Multivariable Control Applications

The inferential models: RVP of debutanizer bottom and iso-pentane of debutanizer
overhead, as discussed above, were implemented on-line as input signals to non-

linear multivariable control system on a debutanizer. These models provided improvement in the performance of closed-loop quality control by making feedback information more timely *i.e.*, improved disturbance rejection. Another advantage of using inferential models in multivariable control applications was to reduce the long time delays caused by the use of analyzers. The use of inferential models as controlled variables improved the performance of closed-loop quality control and the dynamic response of the control system. The inferential models predicting the RVP of platformer feed and iC_5 in the overhead stream of the debutanizer is interfaced with multivariable control in Figure 3.9.

While using the inferential model on-line, it is important to monitor its standard deviation. When its value increases and exceeds a set limit, the validity of the model may be in doubt and its quality is set to "Suspect". It has been shown by Ansari (1992) that the cause may be related to improper functioning of a control loop or a change in process conditions, for example, due to entrainment or fouling. In terms of implementation, the estimated product compositions are used in the same way that analyzer readings are used. The computations finally give a predicted number, which can be used as a process variable in any PID controller or multivariable control.

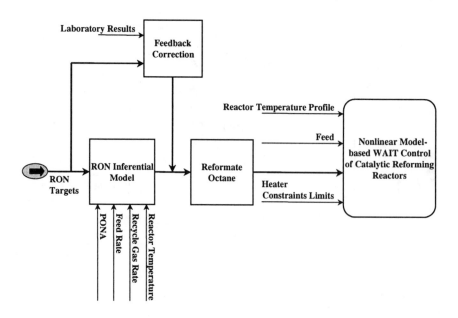

Figure 3.10. System of octane inferential model providing the octane targets to non-linear multivariable control system of the catalytic reforming process

The generic model control (GMC), a non-linear process model-based control strategy developed by Lee and Sullivan (1988) was selected, modified and applied to an industrial debutanizer. The non-linear multivariable process model was directly incorporated into the control algorithm, and interfaced with the inferential models that allowed use of available measurements directly in multivariable control applications. The detail of the non-linear model-based multivariable control is given in Chapter 4. Similar applications of inferential models were carried out in proceeding chapters.

In Chapter 5, the naphtha FBP (top end point) inferential model was developed incorporated in Shell heavy oil fractionator. This model was shown to reduce the long time delays caused by the use of analyzer.

In Chapter 6, implementation of octane inferential model in real-time to a catalytic reforming process demonstrated a successful way to incorporate such type of model in non-linear multivariable control application. Figure 3.10 shows the system of octane inferential model providing the octane targets to non-linear multivariable control system. The model provided an adjustment to the target WAIT for a change in target octane specified. The octane model formed a closed-loop WAIT/octane quality control system and reduced the octane giveaway.

3.6 Benefits of Inferential Models

- The inferential models, apart from being indispensable for the success of the multivariable control applications (linear or non-linear), had additional benefits. The operators do not need to wait for a long time to receive the laboratory results for any adjustment in the process conditions, the inferential model can give them fresh information at any time it is required enabling them to operate more precisely.
- The inferential models are also highly cost effective when compared to installation of new analyzers or a major refurbishment of existing ones. The price of a typical analyzer is estimated to be around US$100,000 excluding installation and maintenance costs. The reformate octane analyzer's cost is US$250,000.
- Inferential models provide operation with real time quality trend information (on-line applications only).
- Better performance of quality control, resulting in less quality give-away and more optimal yield.
- Improvement in the performance of closed loop quality control by making feedback information more timely *i.e.*, better disturbance rejection.
- Inferential models assist operations achieving quality targets and quality constraints using laboratory results as a feedback mechanism.
- Use of inferential models as controlled variables in non-linear multivariable control applications improved the dynamic performance of the controller.

3.7 Conclusions

A wide variety of inferential models were developed in this chapter such as naphtha FBP and kerosene flash point of crude distillation; iso-pentane and RVP of stabilizer and reformate octane number of the catalytic reforming process. These inferential models are shown to perform well in on-line applications on heavy oil fractionator, debutanizer and catalytic reforming process. The on-line applications interfacing with non-linear multivariable control formed a closed-loop quality control system, improving the dynamic performance of multivariable control system by minimizing the long time delays caused by the analyzers. The application helped process operators achieve quality targets, resulting in less quality giveaway and more optimal yield. Operator's acceptance of these models was very high as these models have eliminated the long waiting hours for them to receive the laboratory results for making any adjustment in operating conditions. The prediction of these models was within the repeatability of laboratory results.

These models were developed using correlation-based techniques either obtained from the steady state models of the process or from the experience of plant operation. Other techniques applied recently in development of inferential models such as partial least squares (PLS), non-linear inferential control (NLIC) and neural networks were also highlighted.

The important conclusion here is that when regression techniques alone are used as the only method to predict inferred variables on a distillation columns (fractionator, debutanizer, splitter *etc.*), the results are inferior to the "more rigorous" approach using fundamental chemical engineering principles combined with statistical analysis. The above statement is based on the performance of these models in real-time applications. Inferential models developed based on steady state models with minimum process-model mismatch are robust and predicted more accurate results in the face of varying operating conditions. Also the non-linear inferential models performed well in real-time as compared to the linear models because of their wider range of operations.

CHAPTER 4

NON-LINEAR MODEL-BASED MULTIVARIABLE CONTROL OF A DEBUTANIZER

The debutanizer column control is a major economic incentive to a refinery owing to the importance of the product quality. The non-linear model-based multivariable control on a debutanizer has been shown to provide improved control performance of the top product and the bottom stream qualities using a steady-state process model with approximate dynamics. The control performance is much better than that achieved by the traditional PID-type control system. This chapter outlines the non-linear control method applied to an industrial debutanizer and discusses the use of inferential models to predict the iso-pentane contents in the overhead stream and the Reid vapour pressure in the bottom stream, in non-linear control applications.

4.1 Introduction

The control strategies for chemical processes have been traditionally designed using simple, dynamic, linear, cause-and-effect models to describe the process. Although these models are sufficient for some processes, there are many other processes for which linear models do not provide an accurate basis for good control. The weighted average bed temperature (WABT) control of the platformer's reactors is a typical example of non-linear response of the reactor bed temperature which varies with the variation in the feed composition. Other examples include: the response of the overhead product composition of a distillation column to the reflux flow which changes considerably over its range of operation; the fluid catalytic cracking process exhibits many of the characteristics that are typically present in industrial processing systems, particularly with regards to constraints, interaction and non-linearity.

In this chapter, the development of a non-linear process-model-based control of a debutanizer column is presented. Non-linear model-based control is an integrated approach for process analysis, control and optimization where the same steady state, non-linear, process model is used at each stage. Steady-state non-linear models are selected for two important reasons. Firstly, they are far more abundant than accurate dynamic models in the chemical engineering literature. Secondly, since many chemical processes are non-linear, these models, by their design, preserve the non-linearities of the real process within the equations, of the model.

The use of this type of model to predict the control action required to meet the control objectives can be expected to provide improved performance over simple, linear models. In addition, for multiple-input/multiple-output (MIMO) systems, the model accounts for interaction among the process variables. Therefore, the model can predict the control actions to move all the controlled variables towards setpoint while the single loops may interact with each other.

It is the non-linearities of the process model that provide the improved performance, therefore any controller structure developed must preserve these non-linearities. The generic model control (GMC) structure of Lee and Sullivan (1988) permits the direct use of non-linear dynamic models and, therefore, provides the basic structure of the model-based controller. Since the GMC imbeds a mechanistic process model within the control strategy, it makes sense that this model could also be used for on-line optimization of the process. In this chapter, the GMC structure is further extended to permit the use of steady-state non-linear models in conjunction with estimates of the true process dynamics.

There are a number of control strategies based on using a non-linear approximate model directly for control decisions. Economou *et al.* (1986) extended the internal model control so that non-linear model can be used. This approach is called the non-linear internal model control (NLIMC) and uses an iterative integration of the approximate model for its control law. Riggs and Rhinehart (1988) compared the GMC and the NLIMC for a wide range of exothermic CSTR control problems and exchanger control problems and found that both methods gave essentially the same control performance. They pointed out that the GMC control law is an explicit numerical formulation while the NLIMC is an implicit one so that the GMC is considerably easier to implement and requires less computational effort.

Parrish and Brasilow (1988) used non-linear predictive model control (NLPMC), by taking into consideration the internal model structure but assigned any process/model mismatch to unmeasured disturbances. Bequette (1988) presented a version of NLPMC that used a single-step-ahead control law with continuous model parameter updates. Patwardhan *et al.* (1988) applied the NLPMC for the startup of an open-loop, unstable exothermic CSTR. They considered a maximum limit on the value of the rate of heat addition or the rate of heat removal. The author also applied a version of GMC to the same problem and found that it gave an equivalent performance. Therefore, results to date indicate that there is an insignificant difference between performance of these non-linear process-model-based control (PMBC) methods when the same approximate model is used. In fact, the major difference between the various non-linear PMBC methods is the way in which offset is removed: GMC uses the integral term; NLIMC uses the setpoint bias and NLPMC adjusts the disturbances. These results further suggest that using the approximate model is more important than the way it is applied.

Henson and Seborg (1989) reviewed the field of differential geometric control strategies. By studying a CSTR and a PH control problem, they found that the static methods, of which non-linear PMBC is a subset, provided the best control performance and were relatively insensitive of process/model mismatch.

The non-linear GMC was applied to propylene/propane column by Riggs (1990) based upon the Smoker's equation and compared with the industrial PI controller. The author reported that this controller provided significantly improved control performance over a conventional single loop control system. Cott *et al.* (1989) also reported an industrial based application of the GMC controller on a depropanizer distillation column. The GMC accurately predicted the new steady-state operating conditions, reducing response time and effectively eliminating overshoot of overhead composition. Douglas *et al.* (1994) applied model-based control techniques using GMC on an industrial high purity distillation column operated by Shell Canada.

Recently, Ramachandran (1998) considered neural network steady state model in a process-model-based control application framework to control an industrial distillation column.

The GMC is our method of choice because it is a good approach to implement the non-linear process-model-based control and it has been shown to be more successful in industrial based applications than any other non-linear method. From a design point of view, this approach has several benefits. It eliminates the problems of using different models for different functions and thereby reduces potential optimizer-controller mismatch. It also reduces the effort of implementation, because both the controller and the optimizer can share routine such as the parameter update.

4.2 The Debutanizer Control Strategy

4.2.1 Objective

The objective of the debutanizer control is to control iC_5 in the overheads to less than 0.75% and the RVP of the platformer feed to the target values (100 Kpa) using non-linear model-based control techniques.

4.2.2 Process Description

A simplified flow scheme of the debutanizer is shown in Figure 4.1. The column receives unstabilized naphtha as a feed from the crude distiller. The column fractionates the naphtha such that the light ends are removed from the top and the debutanized naphtha is removed from the bottom and directed to the splitter/platformer section for further processing. In the overhead section, the non-condensed components are removed as the net off-gas to maintain the column pressure while the condensed liquid is directed to the LPG section. A small portion of the condensed liquid from the overhead is used as a reflux to the column whereas the reboiler provides the heat necessary to partially vaporize the debutanizer bottoms liquid before returning it to the column.

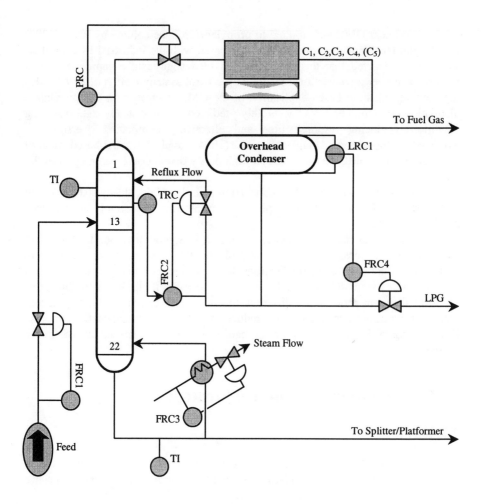

Figure 4.1. Simplified flow scheme of debutanizer showing PID control

4.2.3 Control Proposal

A non-linear model-based multivariable control strategy shown in Figure 4.2 has been suggested and implemented on the debutanizer column. The non-linear multivariable control manipulates the reflux flow controller (FRC2) and the reboiler steam flow controller (FRC3) to control the product quality (i-pentane in LPG) and the bottom quality (RVP of the platformer feed). The disturbance variable is the feed flow to the column and the variation in the feed flow can be handled by the non-linear multivariable control as a feedforward mechanism. The on-line inferential models predicting the i-pentane in the LPG and the RVP of the platformer are used as the controlled variables in this application.

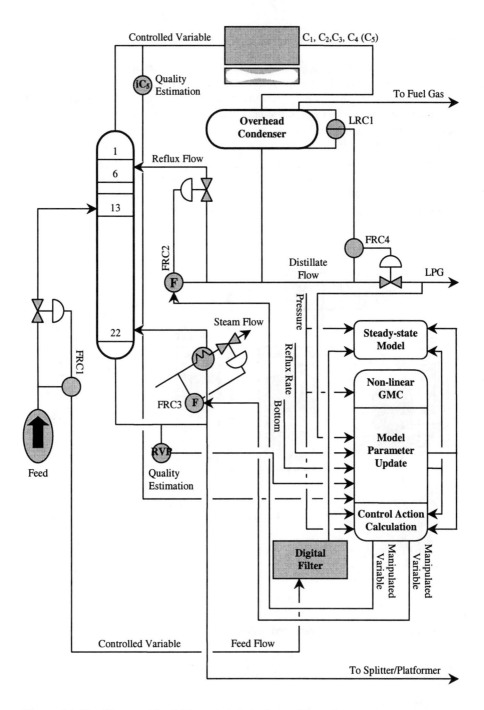

Figure 4.2. Non-linear multivariable control strategies on debutanizer

4.2.4 Hardware Consideration

The control panel was equipped with pneumatic field transmission and conventional pneumatic controllers. To facilitate computer control Taylor MOD30 controllers were installed in the panels wherever a computer-generated setpoint was implemented. The MOD30 digital controller was interfaced with DEC VAX 4000 series process computer through a Taylor communication link. The process computer was used for all the non-linear control applications. A real-time database system was used for the collection of operating data; build up of the database system; and the trend displays for the operators. Figure 4.3 shows the hardware and its connections to the controllers. An important thing to observe in this hardware is that multivariable control does not depend on the hardware used. Also, it does not depend on any kind of distributed control system (DCS).

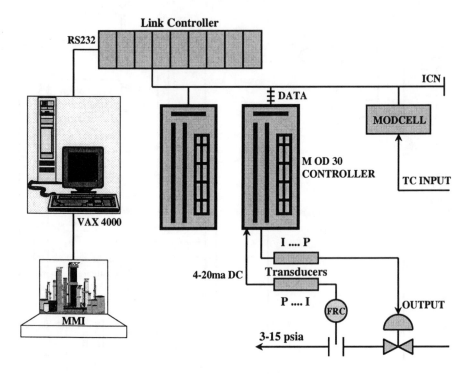

Figure 4.3. Schematic of hardware used in debutanizer application

4.3 The Non-linear GMC Control Law

The GMC control law was developed for a single-input and single output (SISO) system and subsequently extended to the multivariable control application on the debutanizer. Consider a SISO process described by the following approximate model

$$dy/dt = f(y, u, \mathbf{d}, \mathbf{k})$$ (4.1)

where: y = output variable,
        u = manipulated variable,
        $\mathbf{d}$ = vector of the measured disturbances,
        $\mathbf{k}$ = vector of the model parameters.

The control objective is to move the process from y_0 to y_{sp} within a certain time period τ. Now by applying the forward finite difference of the derivative to Equation 4.1, the following expression can be obtained

$$\frac{y_{sp} - y_0}{\tau} = f(y_0, u, d_0, k)$$ (4.2)

Expression 4.2 represents a simple PMBC control law. Each variable is known except u, and Equation 4.2 can be solved to determine u, the control action. τ is a tuning parameter, the smaller the value of τ, the more rapid is the response of the controller.

Since Equation 4.2 uses only an approximate model, a steady-state offset will result from using this control law. Lee and Sullivan (1988) added an integral term (analogous to a PI controller) resulting in a GMC control law which is used to trim the model to account for process/model mismatch.

$$f(y_0, u, \mathbf{d}, \mathbf{k}) + k_1(y_0 - y_{sp}) + k_2 \int (y_0 - y_{sp}) dt = 0$$ (4.3)

where k_1 represents $1/\tau$ in Equation 4.2. The control law given by Equation 4.3 has some desirable characteristics: Throughout the valid operating range of the model, the integral term ensures not only robust process behaviour in the presence of model error, but also the elimination of the steady-state offset. It is a single step law, so that the control calculations are computationally inexpensive.

4.4 GMC Application to Debutanizer

The GMC extension to a multivariable model is straightforward, but the control law for the debutanizer must be modified. A simple approach is to assume that the step response data can be represented by a first-order response. The simple estimate of the response time of the output variables in moving from one steady state to another can be given as:

$$dy / dt \approx T^{-1}(y_{ss} - y)$$ (4.4)

where T is a diagonal matrix of the estimated open loop time constants, and y_{ss} are the ultimate values of the output variables if no further control action is taken. Although this statement may be inaccurate at different operating conditions explained by Lee (1991), the degree of approximation is often sufficient to obtain a good control performance.

Assuming first-order dynamics:

$$dx \,/\, dt = T_x^{-1}(x_{ss} - x) \tag{4.5}$$

$$dy \,/\, dt = T_y^{-1}(y_{ss} - y) \tag{4.6}$$

$$dw \,/\, dt = T_w^{-1}(w_{ss} - w) \tag{4.7}$$

The diagonal elements of the matrix T are "averaged" time constants of the output variables based on step response tests and x_{ss}, y_{ss} and w_{ss} are the solutions of the steady-state approximate model based upon the current values of the measured disturbances (w) and the current values of the state and output variables (x, y). Combining these sets of equations with the GMC control law yields the following modelling equations:

$$x_{ss} = x + k_{1,1}T_x(x_{sp} - x) + k_{2,1}T_x \int_0^t (x_{sp} - x)dt \tag{4.8}$$

$$y_{ss} = y + k_{1,2}T_y(y_{sp} - y) + k_{2,2}T_y \int_0^t (y_{sp} - y)dt \tag{4.9}$$

$$w_{ss} = w + k_{1,3}T_w(w_{sp} - w) + k_{2,3}T_w \int_0^t (w_{sp} - w)dt \tag{4.10}$$

where $k_{1,i}$ (i=1,2,3) = the non-linear model-based control tuning parameters (proportional action) and $k_{2,i}$ (i=1,2,3) = the non-linear model-based control tuning parameters (integral action)

These equations can also be viewed as PI controllers being used to select the target levels. These equations can be directly evaluated to determine x_{ss}, y_{ss} and w_{ss}. Then the boil-up rate, the reflux rate and the disturbance rate (magnitude) are then determined explicitly using the approximate model with the values of x_{ss}, y_{ss} and w_{ss}. The approach used in Equations 4.5, 4.6 and 4.7 assumes that the column exhibits first-order dynamics. Cott et al. (1989) used this approach successfully in applying non-linear PMBC to the control of the depropanizer.

A constrained non-linear optimization structure can be developed for the GMC application to the debutanizer subject to the process constraints present in the system. The method is to accommodate the constraints by establishing the GMC specification curves for the constraint variables and the control variables. The constraint variables can be controlled to the constraint limits as well as to the rate of approach to the constraint limit. Slack variables defining the trajectory departure from the chosen specification curves can be added to the GMC control law for both the control variables and the constraint variables. The solution of the problem then becomes a non-linear constrained optimization, which minimizes a function of the slack variables. The choice of the weighting factors on the slack variables establishes the importance of the process constraints relative to the system setpoint tracking capability. Brown *et al.* (1990) gives the details of this application.

In the debutanizer application, constrained non-linear optimization approach has not been applied as the system presents no input or input movement constraints caused by the physical limitations of the equipment. SetPoint Inc. (1992) imposed the soft constraints on the controlled and the manipulated variables via "InfoPlus", a real-time data management and control system for hydrocarbon processing plant. The "setpoint high and low limits" were the allowable setpoints that InfoPlus implements. The controller did not allow any of the future manipulated variables to move to violate upper or lower limits.

Figure 4.4 illustrates the low and the high constraints imposed on the controlled variable (iC_5). The controller is obliged to keep the controlled variable within the constraint and is allowed to follow any trajectory within these constraints. The controller has the maximum freedom to determine a trajectory that will require minimum manipulated variable movement and will be least sensitive to the modelling errors.

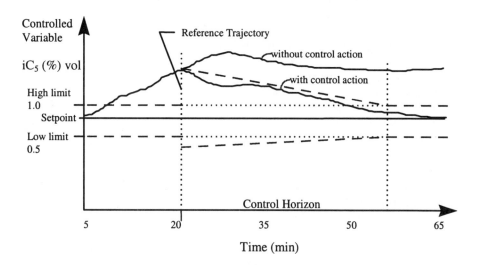

Figure 4.4. Low and high constraints on debutanizer top quality (iC_5)

4.5 Model Development

4.5.1 Steady-State Model Considerations

The model selection procedure is well described by Cott *et al.* (1989), who suggested that the combination of the model selection techniques such as the model accuracy and the model computational effort provide the control system designer a model selection procedure. A number of steady state approximate models is available in the literature, for example: the Smith-Brinkley (S-B) model in Perry and Green (1984); the Douglas-Jafarey-McAvoy (D-J-M) model (1989) and the Fenske-Underwood-Gilliland (F-U-G) model (1984). In general, these models are based on a McCabe-Thiele diagram, and therefore, each assumes that the feed tray has the same composition as the feed. Following the model selection procedure outlined by Cott *et al.* (1989), the Smith-Brinkley model was selected for the debutanizer application. Although other models are of similar accuracy, however, S-B model requires less computational efforts as compared to other models except D-J-M model. Least squares analysis has shown that D-J-M model to be of acceptable accuracy over small ranges of operations. Therefore, in debutanizer column S-B model was selected as it is valid for a wide range of operating conditions.

For the debutanizer column, we assumed a constant relative volatility and a constant molar overflow to enhance the computational efficiency for the model. A steady-state simulation model was developed using actual plant operating conditions. Once the model matched the plant behaviour and predicted the product yields accurately, the GMC S-B model was solved using x_{ss}, y_{ss} and w_{ss} from Equations 4.8 - 4.10 to determine the values of the manipulated variables, the reflux flowrate (R_{Flow}) and the reboiler flowrate (RB_{Flow}).

4.5.2 Development of Inferential Models

The inferential models developed here were based on the correlation derived from the steady-state models for an actual process. The expression may be linear or non-linear, but it was kept as simple as possible to enhance robustness and maintainability. When the models are non-linear, the regression can still be carried out from the process data, because the equations are still linear with respect to the coefficients. The details on inferential models and their applications in non-linear multivariable control system are given in Chapter 3. Here, a brief description of the two inferential models used for top and bottom quality control in debutanizer application is given.

Firstly, the temperatures in the distillation columns are normally pressure compensated, therefore, the pressure-temperature compensation was carried out using the non-linear Clausius-Clapeyron equation given in Perry and Green (1984). It was modified in the following form for this particular application:

$$Tcomp = T_k * B_i / (T_k * LnP + B_i) \qquad (4.11)$$

where Tcomp is a pressure compensated temperature, T_k is the tray temperature in Kelvin and B_i is a Clausius-Clapeyron coefficient.

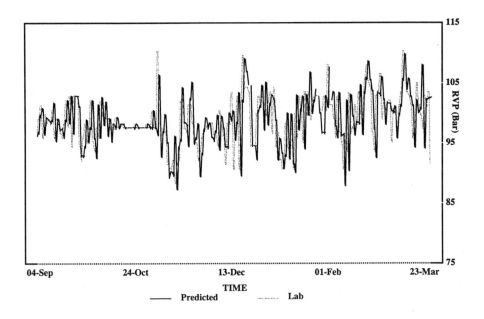

Figure 4.5. Comparison of RVP of platformer feed prediction with laboratory results

4.5.2.1 RVP of Platformer Feed

The correlation developed for the RVP of platformer`feed of a debutanizer is a simple correlation based on the pressure-compensated temperature and 10% ASTM distillation. The laboratory results were used to update or trim the model. The results were compared with laboratory results as shown in Figure 4.5. A normalized distribution was assumed and standard deviations were calculated, based on the observed "laboratory minus predicted" difference. For the most stable plant data, the RVP prediction exhibited a standard deviation of 0.01 bar. The RVP correlation is given by the following equation:

$$RVP = f\,[\,P,\,T,\,10\%\ ASTM,\,\Lambda_1\,] + \Lambda_2 \qquad (4.12)$$

In Equation 4.12, P and T are the column pressure and the bottom tray temperature respectively, Λ_1 is the filter of the first order to reduce the measurement noise and Λ_2 is a bias or a trim factor to update the model from the laboratory quality results. Equation 4.12 was derived from the steady-state model, which receives the scanned values of the pressure and the temperature every two minutes from the real-time data acquisition system. Based on these values the model calculates and predicts the quality of the product. The pressure-compensated temperature obtained from Equation 4.11 incorporates the effect of a pressure variation in the column.

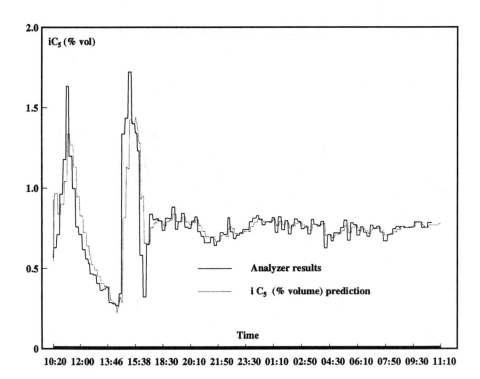

Figure 4.6. Comparison of debutanizer iso-pentane prediction with analyzer results

4.5.2.2 Iso-Pentane of the Debutanizer Overhead

The correlation developed for iso-pentane from a debutanizer overhead is based on the pressure-compensated temperature, the non-ideal gas correlation factor and a bias. The analyzer results were used to update or trim the model. In the case of iC_5, analyzer results were compared with the inferential model predictions and the results are very accurate as shown in Figure 4.6. It is important to mention that the changes predicted by the inferential model were sometimes of the same magnitude as the recognized repeatability of the laboratory test. The on-line analyzer results, when they are available, generally agreed with the predicted values, but they exhibited long time delays. The analyzer measurements were not integrated into the model due to long time delays (15-20 minutes). Instead analyzer results or the laboratory analysis were used to update the model if its standard deviation exceeded the specified limits. While using the inferential model on-line, it is important to monitor its standard deviation. When its value increases and exceeds a set limit, the validity of the model may be in doubt and its quality is set to "Suspect". It has been shown by Ansari (1992) that the cause may be related to the improper functioning of a control loop or a change in process conditions, for example, due to entrainment or fouling.

4.6 Controller Implementation

4.6.1 Controller-Process Interface

The non-linear model-based controller was designed so that it can conveniently interface with the column in a real-time application. The implementation of the model-based controllers usually involves two major steps, a *model parameter update* and a *control action calculation*. The input data for the process was retrieved and stored in a database. Steady-state conditions were identified for model parameterization. The model parameter update provides a form of long-term feedback to the controller allowing it to account for process/model mismatch. The standard form of model parameter update is outlined as follows:

• Check to see if the column is at steady state.
• If it is at steady state, then,
• Measure D, B, RB$_{Flow}$, y, x, R$_{Flow}$, W, z, P for column,
• Perform data reconciliation on the internal and external material balances,
• Solve for the S-B parameters, N and M

In the above parameter update procedure, D is the distillate, RB$_{Flow}$ is the reboiler flow, B is the bottom vapour return to column, y and x are the overhead and bottom compositions, R$_{Flow}$ is the reflux flow, W is the feed flowrate, z is the feed composition, P is the column pressure, N is the total number of stages and M is the number of stages below the feed tray.

The data reconciliation step eliminates the measurement errors from the flowrates and provides the consistent internal and external material balances. Given this constant data, the S-B model can be inverted to solve for the parameters, N and M. An algorithm for data reconciliation is given in Crowe (1986). The algorithm to solve the GMC controller using S-B model is given below:

Algorithm 4.1. GMC controller using S-B model

Step 1. Measure W, z, P from column and put through the digital filter for the lack of dynamic information in the S-B model.
Step 2. Get the model parameter values N and M from the model parameter update,
Step 3. Using the filtered data, solve GMC control law given by Equations 4.8 - 4.10.
Step 4. Solve for the values of manipulated variables R$_{Flow}$ and RB$_{Flow}$ needed to obtain x$_{ss}$ and y$_{ss}$ using the S-B model equations.

A combined structure of S-B GMC controller with information flow diagram for the controller is given in Figure 4.7 and the equations used related to implementation of S-B GMC controller are given in Appendix B.

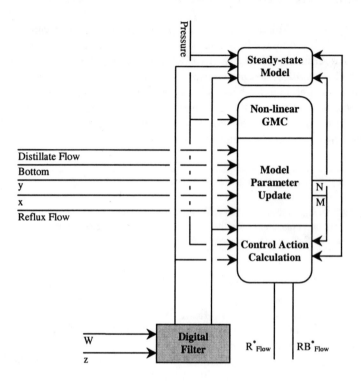

Figure 4.7. Structure of combined S-B GMC controller: signal flow diagram

Following the model parameter update procedure, the model was checked for the closure of material and energy balances. The steady-state validation test was run at a frequency of once per 10 minutes. The feed flow rate was passed through a digital filter for feedforward action of the non-linear controller. Using the filtered data the GMC control law given by the Equations 4.8 - 4.10 was obtained.

The dynamic matrix optimization (DMO) package was used to build up and solve the S-B GMC controller. This package consists of an efficient non-linear equation solver/optimizer "engine", coupled with robust, detailed, mechanistic process simulation models, an on-line interface serves several functions, including steady state detection, validity checking of data, scheduling of tasks and ramping of the optimization solution to the process control computer where the controllers execute.

The standard model of stabilizer from the process model library (PML) of DMO was used and modified to incorporate the S-B steady state model. The S-B GMC controller then solved using the non-linear equation solving techniques in DMO. An alternate approach to solving the resulting set of differential equations is to constitute an initial value problems and to integrate these equations by using the Gear type integration package (LSODE) (Hindmarsh, 1980) which considers the banded nature of the Jacobian of the set of differential equations. This approach was applied by Riggs (1990) to solve the non-linear control problem for a propylene sidestream draw column.

The process model was used to predict the manipulated and controlled variable values and operating conditions under different feed changes and unstabilized naphtha feed compositions. Two cases were studied at different operating conditions (feed rate changes) and the results are given in Tables 4.1 and 4.2. The laboratory analysis of the feed has shown no changes in the feed compositions, therefore feed composition (z) was not considered in the case-studies.

Table 4.1. The estimate of manipulated and controlled variables. Case1: Higher unstabilized naphtha feed rate (150 m^3/hr).

Operation variables	Model Results	Plant Test	Relative Error
Reboiler Flow (RB$_{Flow}$) Reflux Flow (R$_{Flow}$) Distillate Flow, LPG (D) Bottom Flow (B)	90 m^3/hr 80 m^3/hr 10 m^3/hr 128 m^3/hr	93 m^3/hr 84 m^3/hr 9.7 m^3/hr 130 m^3/hr	3.3% 4.8% 3.1% 1.6%
Stabilizer iso-pentane iC$_5$, Vol %	1.50	1.42	5.6%
Stabilizer Reid Vapour Pressure RVP, kPa	105	110	4.8%

Table 4.2. The estimate of manipulated and controlled variables. Case2: Lower unstabilized naphtha feed rate.

Operation variables	Model Results	Plant Test	Relative Error
Reboiler Flow (RB$_{Flow}$) Reflux Flow (R$_{Flow}$) Distillate Flow, LPG (D) Bottom Flow (B)	80 m^3/hr 73 m^3/hr 7 m^3/hr 115 m^3/hr	82 m^3/hr 76 m^3/hr 7.2 m^3/hr 118 m^3/hr	2.5% 3.9% 2.8% 2.6%
Iso-pentane iC$_5$, Vol %	1.2	1.15	4.3%
Stabilizer Reid Vapour Pressure RVP, kPa	102	106	3.9%

Table 4.1 shows that the maximum relative errors in the estimation of the manipulated variables: reboiler flow and reflux flow between the model prediction and plant are 3.3% and 4.8% respectively. At lower feed rate (Table 4.2), this error is reduced to 2.5% and 3.9%. At higher feed rate, the relative errors in the prediction of controlled variables by the model and plant test is higher (5.6% in iC$_5$ and 4.8% in RVP) as compared to the relative error at the lower feed rate (4.3% in iC$_5$ and 3.9% in RVP). This difference in the relative errors can be explained by the fact that at higher feed rate, the reflux flow operates at maximum capacity and

reboiler is at a maximum heat duty. The cooling capacity of the overhead condenser may be a constraint limiting the iC_5 to meet the target specification of 0.75% volume and RVP target at 100 kPa. This case was further investigated during the implementation of non-linear multivariable control applications on debutanizer.

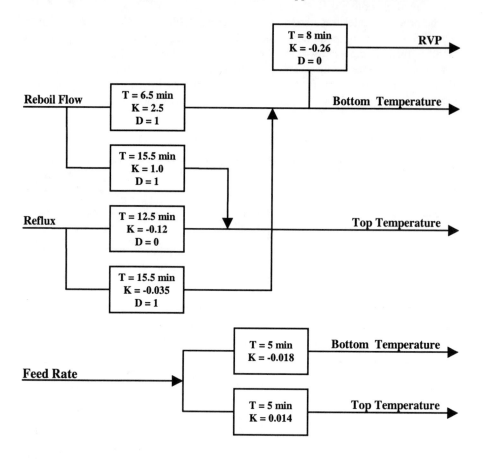

Figure 4.8. Results from the process identification test of debutanizer

4.6.2 Non-linear Controller Tuning

The time constants of the debutanizer column were estimated from the open-loop test of the reflux and the steam flow rates (manipulated variables) and the feed flow rate (disturbance variable). The results of the process identification of the debutanizer are shown in Figure 4.8. It can be observed from this figure that the process models obtained from the process identification tests of debutanizer are not dominated by the dead time. In fact a *reflux - top temperature model* shows no dead time. Similarly, the *reboiler flow - RVP model* also gives no dead time. This shows that the system has a very fast dynamics. Since the dead time on the debutanizer is not significant, therefore it was not considered in the controller structure.

The largest time constant $\tau = 15.5$ minutes is between the *reboiler flow-top temperature model*. Based on this process identification test τ was selected for GMC parameters given in Table 4.3. These time constants are the diagonal elements of the output variables used in Equations 4.5 - 4.7.

Table 4.3. Setpoint and GMC parameters

Output	Setpoints	GMC Parameters	
		ξ	τ (min)
iC_5	0.75%	3	16
RVP	100kpa	3	16

The sampling frequency was taken as once per minute. For each output variable in the Table 4.3, the GMC controller parameters K_{1i} and K_{2i} (the ith diagonal elements of K_{11} and K_{21}) from Equations 4.8 - 4.10 were calculated by the following relationships:

$$K_{1i} = \frac{2\xi_i}{\tau_i} \tag{4.13}$$

$$K_{2i} = \frac{1}{\tau_i^2} \tag{4.14}$$

These two tuning parameters bear a close resemblance to the tuning constants of a classical PI controller: K_{11} is like the PI controller gain, and K_{21} is like the product of the gain and reset. τ_i was estimated from the process identification test (plant test runs) and ξ_i was selected from the GMC performance specification curve (Lee and Sullivan,1988). This pair of parameters has explicit effects on the closed-loop response of output variable. The parameter ξ_i determines the shape of the closed-loop response while τ_i determines the speed of the response (large τ_i means slow response).

The control application was first turned on and tested in an open-loop mode (*i.e.*, the outputs were suppressed). Open loop testing includes verifying the process inputs and the checking of the controller move sizes and direction. The controls were then tested in the closed-loop mode using tuning constants gathered from the simulation model and the step response test. The closed-loop testing was performed within a narrow operating range until confidence was gained. During this period tuning adjustments were made as necessary. Initially the controllers were left for a few hours at a time, as each shift was trained on the new application and confidence was gained, the controllers remained on permanently.

4.7 Results and Discussions

Figure 4.9 shows a good performance of non-linear multivariable control application on a debutanizer top quality. When the column is on a PID control, the iso-pentane in the top stream varies between 0.6 to 1.7% vol. from the target setpoint of 0.75% vol. whereas the non-linear model-based control kept the iso-pentane at the target values. It can also be observed from Figure 4.10 that the RVP of the Platformer feed deviates by 15-20 kpa from the target setpoint of 100 kpa. on a PID control and after the implementation of the non-linear control, the deviation in the RVP was reduced to 0-5 kpa.

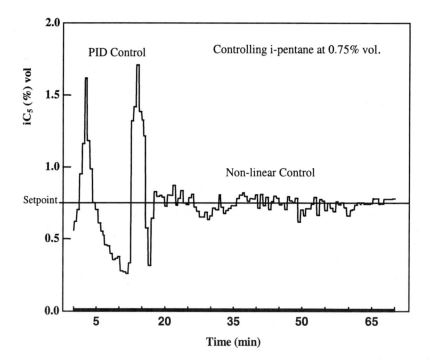

Figure 4.9. Non-linear model-based control on debutanizer top quality

It was also observed, during the operation of the debutanizer, that any disturbance in the feed flow was not taken care of by the PID control system, which influence the top product and the bottom stream specifications. The demand on this column was to meet both the top and the bottom quality targets in the face of feed disturbances. Non-linear model-based control has been shown to satisfy these demands. The Figures 4.9 and 4.10 demonstrate improved control performance over the traditional PID-type control system.

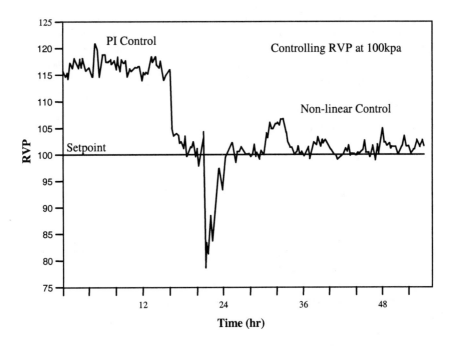

Figure 4.10. Non-linear model-based control on debutanizer bottom quality

Long-time delays (15-20 min) were observed by using the analyzer in the non-linear control application. Inferential models, based on fast and continuously available temperature and pressure were developed to reduce the negative impact of these sample intervals and the time delays and the laboratory results were used to update the quality of predictions. For a safe operation, when the non-linear controller was turned off, the PID controller setpoints hold their last values (bumpless transfer) by allowing the controllers to "track" the process.

4.8 Cost/Benefit Analysis

Economic justification of advanced process control projects can be carried out by following one of the three methods outlined:

1. Estimation based on experience.
2. After-control minus before-control audits.
3. Estimates based on analysis of before-control data.

Estimates based on experience call for a broad experience base and are usually calculated by some form of scale-up procedure. Such estimates often lack sufficient credibility to satisfy top management.

After-controls minus before-control audits are convincing, since they are based on actual data from the controlled unit, and calculate bottom-line profit. The main problem with this approach is that it cannot be used until the project is completed.

Estimates based on analysis of before-control data are appealing as they pertain to the specific unit in question and can be completed prior to the project.

In the debutanizer application, the method "after-controls minus before-control audits", for calculating the benefits was preferred as the project was completed and the data was available before and after the control.

4.8.1 Benefits Calculations

From Figure 4.10 it can be observed that iC_5 specification is set at 0.75% vol. Under PID control the average deviation of iC_5 is 1.15% vol. Non-linear control reduces the iC_5 variation to 0.2% vol. The non-linear controller improves the deviation to 0.95% vol. This improvement in deviation of iC_5 reduces total butane (C_4) by 6% vol. In the bottom of debutanizer which improves the hysomer (downstream unit) feed by 20 tons per day. That means conversion of extra 20 tons per day nC_5 to iC_5.

The conversion of nC_5 to iC_5 = 25 $/ton
Consider stream days per year   = 330

Benefits = extra feed x Δ_{cost} (nC_5 - iC_5) x nr. of stream days
                    = 20 tons/day x 25 $/ton x 330 days/year
                    = Australian $165,000/year
Cost to implement the project = A$100,000
(as per the equipment shown in Figure 4.3)

4.9 Conclusions

The non-linear model-based control on a debutanizer has been shown to provide improved control performance of the top product and the bottom stream qualities using a steady-state process model with approximate dynamics. The control performance is much better than that achieved by the traditional PID-type control system.

The use of inferential models as controlled variables in the non-linear model-based control has been very rewarding both in terms of reducing the long-time delays and enhancing the potential of the non-linear control to keep the product qualities close to their setpoints. They also assist the operators to drive the process closer to the optimum. The benefits of the non-linear control were very substantial, for example, on-line times have been high for this application, which indicates that the operators are satisfied with the performance of the controller.

Other interesting feature of this non-linear multivariable control application is that it was implemented on a hardware system without distributed control functionality. This suggests that a distributed control system is not a prerequisite hardware for multivariable control implementation.

CHAPTER 5

NON-LINEAR MODEL-BASED MULTIVARIABLE CONTROL OF A CRUDE DISTILLATION PROCESS

In this chapter, a multivariable control problem is solved by formulating the non-linear constrained optimization strategy for a crude distillation process. The heavy oil fractionator problem proposed by Shell Oil was selected for this work. The presence of hard constraints in the optimization problem makes the overall system non-linear even though the process dynamics are assumed linear. The heavy oil fractionator control problem embodies the key elements of all control problems such as multiple objective criteria, inequality and equality constraints and model uncertainty with first-order differential equations and dead time.

In this chapter, a method for solving the Shell control problem is presented. The method uses the non-linear model-based controller, which considers the model uncertainty explicitly. The method is based on formulating the constrained non-linear optimization (NLP) programme that optimizes performance objectives subject to constraints. Within this framework all the important problem features, which include the combination of both economic and control performance objectives, the direct use of uncertain model presentation, and the presence of constraints on manipulated, controlled and associated variables.

A dynamic model of a heavy oil fractionator introduced by Prett and Morari (1987) was used for dynamic analysis of the plant and non-linear multivariable control system. The model was built in MATLAB®/SIMULINK® and the results were tested and compared with the results obtained from non-linear model-based multivariable control techniques applied to heavy oil fractionator. The non-linear control strategies were implemented in real-time to 120,000 barrels/day crude fractionator.

5.1 Introduction

The crude distillation process is the most important primary process in a petroleum refinery, because of the large feed flow and the amount of energy consumed. The product specifications basically follow the market demands and the economic objective of the crude unit is to keep the products as close to the specifications as possible. To exceed the specifications is less critical than to violate them. Any violation on the product specifications leads to an expensive reprocessing or to degrade the stream to a less valuable pool. The experienced operator keeps a safety margin from the market specifications to prevent any violations during transient conditions. The operating economics of a crude unit generally address the issues of minimizing the "give-away" of higher value product into lower valued streams.

The variability of crude unit feedstock is the major motivation factor for implementing advanced computer controls. It is not unusual for the crude to be switched over on a weekly basis and for the crude stocks to have varying physical characteristics. The objective of advanced controls is to maintain constant product qualities in spite of changing feedstocks and evolving operating conditions. When compared with manual operations, these advanced controls can translate into savings worth up to several million dollars per year, depending on the size and complexity of the crude unit.

The application of model-predictive multivariable control to crude fractionators has already been reported in the literature. Muske *et al.* (1991) describes the application of the IDCOM.M package to a crude tower of the refinery of Sakai in Japan. The process dynamics given by Muske *et al.* (1991) shows that the yield of a side draw does not particularly affect the cut-points of the products above that draw and a triangular structure is obtained. This can be explained, since any change in the flow of the side draws is compensated by the overhead temperature controller that changes the reflux flow and tends to keep the ratio L/V constant. The disadvantage of this control configuration is that the overhead temperature controller can not work at its saturation limit. Other applications of model-based multivariable control on crude unit are reported by Cutler and Finlayson (1988), O'Conner *et al.* (1991) and Hsie (1989). Magalhaes and Odloak (1995) described the industrial application of a multivariable predictive controller using linear dynamic matrix control (LDMC) developed by Morshedi *et al.* (1985). The developed control algorithm showed that the LDMC formulation was adequate to large systems where all the variables were controlled by range, and the manipulated variables have min/max constraints.

This chapter provides details of application of the constraint non-linear multivariable control and solves a non-linear optimization problem on a heavy oil fractionator, a problem proposed by Prett and Morari (1987). The main contribution of this work is to integrate dynamic process model suggested by Shell Oil and to develop a non-linear constrained optimization algorithm and its application to heavy oil fractionator with multiple constraints and model uncertainties. The model was built in MATLAB®/SIMULINK® and the constrained non-linear multivariable control results were tested and compared with the results obtained from the linear model-predictive control method. The non-linear control strategies were implemented in real-time to 120,000 barrels/day crude fractionator.

5.2 Crude Distillation Process Control Overview

5.2.1 Problem Description

An overview of the heavy oil fractionator is shown in Figure 5.1. This figure shows that the fractionator has three product draws and three side circulating loops.

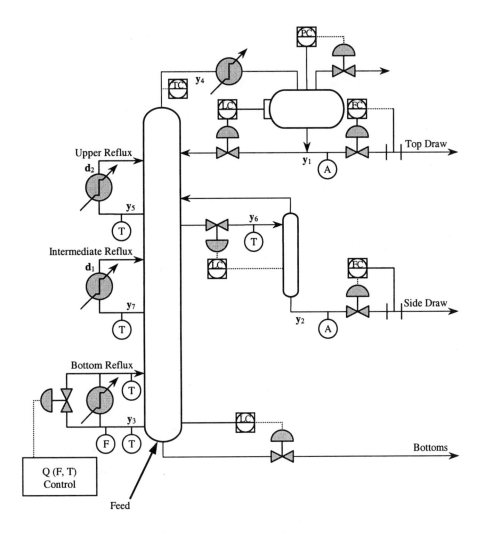

Figure 5.1. An overview of crude distillation process (Shell heavy oil fractionator)

The heat requirement of the column enters with the feed, which is a gaseous stream. Product specifications for the top and side draws are determined by economics and operating requirements. There is no product specification for the bottom draw, but there is an operating constraint on the temperature in the lower part of the column. The three circulating loops remove heat to achieve the desired product separation. The heat exchangers in these loops reboil columns in other parts of the plant. Therefore they have varying heat duty requirements. The bottom loop has an enthalpy controller which regulates heat removal in the loop by adjusting steam make. Its heat duty can be used as a manipulated variable to control the column. The heat duties of the other two loops act as disturbances to the column.

The relevant information regarding the Shell control problem is stated in the following sections.

5.2.2 Control Objectives and Constraints

The Shell heavy oil fractionator problem has seven outputs and five inputs. Three of the inputs can be used for control purposes and two are considered to be unmeasurable disturbances. Of the five measurable outputs, only three are involved in either the control objectives or the constraints; the other four outputs are considered to be auxiliary. Multiple control objectives given in the problem description address such issues as steady state offset, disturbance rejection, optimization of steam generation and obtaining a desired closed loop settling time. Hard constraints are given for maximum and minimum bounds on all of the manipulated variables and top end point as well as a maximum bound on the magnitude of change in manipulated variables. In addition, minimum values are given for the bottom reflux draw temperature and sampling time.

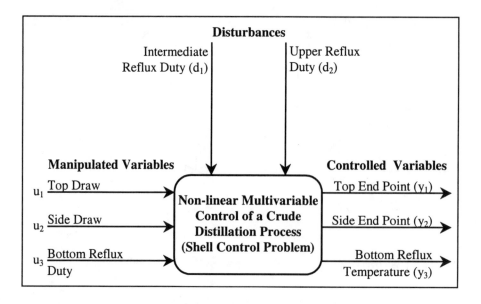

Figure 5.2. Non-linear model-based multivariable control problem schematic for crude distillation process (heavy oil fractionator)

As no constraint and control objectives were specified on the top temperature, upper reflux temperature, side draws temperature and intermediate reflux temperature, these variables have not been considered further. The problem was reduced to a system of five inputs with two disturbances and three outputs as shown in Figure 5.2. This figure gives schematic of control problem identifying disturbances, manipulated variables, and controlled variables.

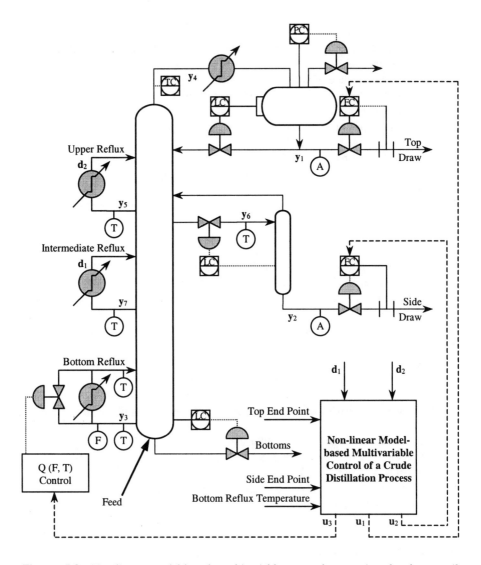

Figure 5.3. Non-linear model-based multivariable control strategies for heavy oil fractionator

A non-linear model-based multivariable control strategies using the features of model-based control suggested by Lee and Sullivan (1988) have been applied to the heavy oil fractionator. This non-linear model-based technique has the ability to handle the multivariable interactions and constraints, its ease of configuration and implementation, adaptability and its robustness. Although it can be applied to complete crude distillation process, including the crude column and stabilizer sections, this chapter will focus only on the Shell heavy oil fractionator, a problem proposed by Prett and Morari (1987). Figure 5.3 shows the non-linear model-based control strategies for heavy oil fractionator.

5.2.2.1 Control Objectives
1. Maintain the top end point and the side end point at specification (0.00 ± 0.005 at steady state).
2. Achieve steady-state offset in TEP and SEP less than 0.5 for arbitrary IRD and URD step disturbances of magnitude less than 0.5 in the event of end point analyzer failure.
3. Maximize steam make in the steam generator in the bottom circulating reflux, *i.e.*, minimize the bottom reflux duty.
4. Reject the disturbances entering the column from the upper and intermediate refluxes due to changes in the heat duty requirements from other columns. Upper and intermediate duties range between -0.5 and +0.5.
5. Keep the closed loop speed of response between 0.8 and 1.25 of the open loop process bandwidth.

5.2.2.2 Control Constraints
1. All draws must be within high and low limits of 0.5 and –0.5.
2. The bottom reflux duty is constrained within the limits of 0.5 and –0.5.
3. All manipulated variables have maximum move size limitations of magnitude 0.05 per minute.
4. The bottom reflux draw temperature is constrained within the limits of 0.5 and –0.5.
5. The top end point and the side end point should be maintained within the maximum and minimum values of 0.5 and –0.5.

5.2.3 Dynamic Model of Heavy Oil Fractionator with Uncertainty

The process model for the heavy oil fractionator is presented as a 7x5 transfer function matrix where each element is given by a first-order transfer function with dead time. The process model is given in Table 5.1. The problem features linear models with uncertain steady-state gains, along with a correlation description. These uncertainties are present on all system transfer functions. Five prototype test cases with different disturbance direction and gain uncertainty directions are specified. Table 5.2 gives uncertainties in the gains of the process model. The detail of five prototype cases is also provided to completely define the problem.

The uncertain model parameters ε_i, i =1,..., 5 are independent and vary between ± 1, *i.e.*, the actual plant may correspond to any combinations of ε_i as long as they are all less than 1 in magnitude. It is also understood that the true plant may change during operation in the sense that the ε_i need not be fixed (although unknown) but may be time varying. The gain uncertainties given in the problem statement appear to be an attempt to incorporate the non-linear behaviour of the column into a linear model framework. The actual gain changes are probably due to high product purity effects. An alternative approach would be to model the column with a low-order, non-linear model that describes the actual separation profile in the column (Marquardt, 1986).

Table 5.1. Heavy oil fractionator process dynamic model $\dfrac{K e^{-\theta s}}{\tau s + 1}$, units for θ, τ are minutes

	Top Draw u_1	Side Draw u_2	Bottom Reflux Duty u_3	Inter. Reflux Duty d_1	Upper Reflux Duty d_2
Top End Point Y_1	$K = 4.05$ $\tau = 50$ $\theta = 27$	$K = 1.77$ $\tau = 60$ $\theta = 28$	$K = 5.88$ $\tau = 50$ $\theta = 27$	$K = 1.20$ $\tau = 45$ $\theta = 27$	$K = 1.44$ $\tau = 40$ $\theta = 27$
Side End Point Y_2	$K = 5.39$ $\tau = 50$ $\theta = 18$	$K = 5.72$ $\tau = 60$ $\theta = 14$	$K = 6.90$ $\tau = 40$ $\theta = 15$	$K = 1.52$ $\tau = 25$ $\theta = 15$	$K = 1.83$ $\tau = 20$ $\theta = 15$
Bottom Reflux Temp. y_3	$K = 4.38$ $\tau = 33$ $\theta = 20$	$K = 4.42$ $\tau = 44$ $\theta = 22$	$K = 7.20$ $\tau = 19$ $\theta = 0$	$K = 1.14$ $\tau = 27$ $\theta = 0$	$K = 1.26$ $\tau = 32$ $\theta = 0$
Top Temp. Y_4	$K = 3.66$ $\tau = 9$ $\theta = 2$	$K = 1.65$ $\tau = 30$ $\theta = 20$	$K = 5.53$ $\tau = 40$ $\theta = 2$	$K = 1.16$ $\tau = 11$ $\theta = 0$	$K = 1.27$ $\tau = 6$ $\theta = 0$
Upper Reflux Temp. y_5	$K = 5.92$ $\tau = 12$ $\theta = 11$	$K = 2.54$ $\tau = 27$ $\theta = 12$	$K = 8.10$ $\tau = 20$ $\theta = 2$	$K = 1.73$ $\tau = 5$ $\theta = 0$	$K = 1.79$ $\tau = 19$ $\theta = 0$
Side Draw Temp. Y_6	$K = 4.13$ $\tau = 8$ $\theta = 5$	$K = 2.38$ $\tau = 19$ $\theta = 7$	$K = 6.23$ $\tau = 10$ $\theta = 2$	$K = 1.31$ $\tau = 2$ $\theta = 0$	$K = 1.26$ $\tau = 22$ $\theta = 0$
Inter. Reflux Temp. Y_7	$K = 4.06$ $\tau = 13$ $\theta = 8$	$K = 4.18$ $\tau = 33$ $\theta = 4$	$K = 6.53$ $\tau = 9$ $\theta = 1$	$K = 1.19$ $\tau = 19$ $\theta = 0$	$K = 1.17$ $\tau = 24$ $\theta = 0$

Table 5.2. Uncertainties in the gains of the model $-1 \le \varepsilon_i \le 1$; $i = 1, 2, 3, 4, 5$

	Top Draw	Side Draw	Bottom Reflux Duty	Inter. Reflux Duty	Upper Reflux Duty
	u_1	u_2	u_3	d_1	d_2
Top End Point y_1	$4.05+2.11\varepsilon_1$	$1.77+0.39\varepsilon_2$	$5.88+0.59\varepsilon_3$	$1.20+0.12\varepsilon_4$	$1.44+0.16\varepsilon_5$
Side End Point y_2	$5.39+3.29\varepsilon_1$	$5.72+0.57\varepsilon_2$	$6.90+0.89\varepsilon_3$	$1.52+0.13\varepsilon_4$	$1.83+0.13\varepsilon_5$
Bott. Reflux Temp. y_3	$4.38+3.11\varepsilon_1$	$4.42+0.73\varepsilon_2$	$7.20+1.33\varepsilon_3$	$1.14+0.18\varepsilon_4$	$1.26+0.18\varepsilon_5$
Top Temp. Y_4	$3.66+2.29\varepsilon_1$	$1.65+0.35\varepsilon_2$	$5.53+0.67\varepsilon_3$	$1.16+0.08\varepsilon_4$	$1.27+0.08\varepsilon_5$
Upper Reflux T. y_5	$5.92+2.34\varepsilon_1$	$2.54+0.24\varepsilon_2$	$8.10+0.32\varepsilon_3$	$1.73+0.02\varepsilon_4$	$1.79+0.04\varepsilon_5$
Side Draw Temp. y_6	$4.13+1.71\varepsilon_1$	$2.83+0.93\varepsilon_2$	$6.23+0.30\varepsilon_3$	$1.31+0.03\varepsilon_4$	$1.26+0.02\varepsilon_5$
Inter. Reflux Temp. y_7	$4.06+2.39\varepsilon_1$	$4.18+0.35\varepsilon_2$	$6.53+0.72\varepsilon_3$	$1.19+0.08\varepsilon_4$	$1.17+0.01\varepsilon_5$

Prototype test cases given in the Shell control problem description (Prett and Morari, 1987) suggest that the required controller should satisfy the control objectives without violating the control constraints for the following plants within the uncertainty set. Assuming all inputs and outputs are initially at zero; magnitude for the upper and intermediate reflux duty step changes are given below:

1. $\varepsilon_1 = \varepsilon_2 = \varepsilon_3 = \varepsilon_4 = \varepsilon_5 = 0$, upper reflux duty (URD) = 0.5, intermediate reflux duty (IRD) = 0.5.
2. $\varepsilon_1 = \varepsilon_2 = \varepsilon_3 = \varepsilon_4 = \varepsilon_5 = 0$,   URD = - 0.5, IRD = - 0.5.
3. $\varepsilon_1 = \varepsilon_3 = \varepsilon_4 = \varepsilon_5 = 1$, $\varepsilon_2 = -1$,   URD = - 0.5, IRD = - 0.5.
4. $\varepsilon_1 = \varepsilon_2 = \varepsilon_3 = \varepsilon_4 = \varepsilon_5 = 1$,   URD = 0.5, IRD = - 0.5.
5. $\varepsilon_1 = -1$, $\varepsilon_2 = 1$, $\varepsilon_3 = \varepsilon_4 = \varepsilon_5 = 0$,   URD = - 0.5, IRD = - 0.5.

Four cases are tested. They are:
1. Positive step changes 0.5 in both disturbances (URD and IRD).
2. Negative step changes - 0.5 in both disturbances (URD and IRD).

3. Setpoint change of top end point from 0.0 to 0.5.
4. Setpoint change of side end point from 0.0 to 0.5.

The parameters of the control system are shown in Table 5.3 in Section 5.6 along with some details on the selection of these parameters.

5.3 Non-linear Control Algorithm for Fractionator

For non-linear control algorithm formulation, the system was simplified with regard to control objectives and constraints as shown in the schematic of the control problem in Figure 5.2. As no constraint and control objectives were specified on the top temperature, upper reflux temperature, side draw temperature and intermediate reflux temperature, these variables have not been considered further.

Select: u_1, u_2, u_3, λ^+_p, λ^-_p, λ^+_c, λ^-_c

where u_1 is the top draw rate, u_2 is the side draw rate and u_3 is the bottom reflux duty which is required to be minimized.

To Minimize:

$$J = \mathbf{W}^-_{pi}\, \lambda^-_{pi\,(i=1,..4)} + \mathbf{W}^+_{pi}\, \lambda^+_{pi} + \mathbf{W}^-_{ci}\, \lambda^-_{ci\,(i=1,..6)} + \mathbf{W}^+_{ci}\lambda^+_{ci} + \Delta\, \mathbf{u}_{i\,(i=1,2,3)}$$
$$(5.1)$$

where $\mathbf{W}$ are weighting matrices with elements such that $w_{ii} \geq 0$, $w_{ij} = 0$, $i \neq j$ and i and j denote weighting factors on the ith and jth slack variables. $\mathbf{u}_i$ are the manipulated variables as defined above. In addition, the positive values of λ^-_p or λ^+_p represent the difference between output response and the GMC reference trajectory. This difference is a measurement of the control performance degradation. Positive values of λ^-_c or λ^+_c, are the differences between the actual rates of changes of the constraints beyond the specified maximum rate of approach towards the constraint bounds. The larger they are, the more likely that the constraints will exceed their bounds in the future. Thus they are the measurement of the potential constraint violation. All of these λs are desired to be minimized.

The slack variable weighting factors, $\mathbf{W}$, reflect the relative importance of the outputs and constraints. Thus, if the constraint control is more important than the quality control, then:

$$\mathbf{W}^+_c, \mathbf{W}^-_c \gg \mathbf{W}^+_p, \mathbf{W}^-_p \qquad\qquad (5.2)$$

Conversely, if the quality control is more important than the constraint control, then:

$$\mathbf{W}^+_p, \mathbf{W}^-_p \gg \mathbf{W}^+_c, \mathbf{W}^-_c \qquad\qquad (5.3)$$

for which the functional constraints virtually have no effect on the closed-loop responses of the system.

Subject to:

$$\frac{dy_1}{dt} + \lambda^+_{p_1} - \lambda^-_{p_2} = k_{11}(y_{1sp} - y) + k_{21} \int_0^t (y_{1sp} - y)\, dt$$

(5.4)

Equation for the top end point

$$\frac{dy_2}{dt} + \lambda^+_{p_3} - \lambda^-_{p_4} = k_{12}(y_{2sp} - y) + k_{22} \int_0^t (y_{2sp} - y)\, dt$$

(5.5)

Equation for the side end point

$$\frac{dy_3}{dt} + \lambda^+_{p_5} - \lambda^-_{p_6} = k_{13}(y_{3sp} - y) + k_{23} \int_0^t (y_{3sp} - y)\, dt$$

(5.6)

Equation for the bottom reflux temperature

Since the bottom reflux temperature is an associated variable and not an important variable to control to specification, the Equation 5.6 was not used in the control algorithm. This variable was kept under lower constraint limit as specified in the control problem.

$$\frac{dy_1}{dt} - \lambda^-_{c_1} \le K^U_{1C}(y_{1U} - y)$$

(5.7)

$$\frac{dy_1}{dt} + \lambda^-_{c_2} \le K^L_{1C}(y_1 - y_{1L})$$

(5.8)

$$-0.5 \le y_1 \le +0.5$$

Equations for the top end point constraints

$$\frac{dy_2}{dt} - \lambda^-_{c_3} \le K^U_{2C}(y_{2U} - y)$$

(5.9)

$$\frac{dy_2}{dt} + \lambda^-_{c_4} \le K^L_{2C}(y_2 - y_{2L})$$

(5.10)

Equations for the side end point constraints

$$\frac{dy_3}{dt} - \lambda\bar{}_{c5} \le K^U_{3C} (y_{3U} - y) \tag{5.11}$$

$$\frac{dy_3}{dt} + \lambda\bar{}_{c6} \le K^L_{3C} (y_3 - y_{3L}) \tag{5.12}$$

$$-0.5 \le y_{3L} \text{ (only low constraint)}$$

Equations for bottom reflux temperature constraints

$$\mathbf{u}_{iL} \le \mathbf{u}_i (t) \le \mathbf{u}_{iU}, \qquad i= 1,2,3 \text{ (input constraint)} \tag{5.13}$$

$$-0.5 \le \mathbf{u}_i (t) \le +0.5, \qquad i= 1,2,3$$

$$\Delta\mathbf{u}_{iL} \le \Delta\mathbf{u}_i (t +\Delta t) - \Delta\mathbf{u}_i (t) \le \Delta\mathbf{u}_{iU}, \text{ (input movement constraint)} \tag{5.14}$$

$$-0.05 \Delta t \le \Delta\mathbf{u}_i = \le +0.05 \Delta t$$

where sampling time Δt is equal to 5 minutes.

$$\lambda^+_p \ge 0, \lambda^-_p \ge 0, \lambda^+_c \ge 0, \lambda^-_c \ge 0 \tag{5.15}$$

The optimization problem can be solved using a non-linear constrained optimization problem algorithm. The form of optimization problem is well structured since a slack variable is added to each control law equation to ensure that a solution to the set of equations does exist. If the control is implemented at a reasonable frequency, the solution of the NLP is very fast since the current control settings and slack variables provide a good initial estimate of the solution vector. Brown *et al.* (1990) have applied this control algorithm to two non-linear simulated systems. The first system was a forced circulation single-stage evaporator and the second system was a stirred tank reactor from the paper of Li and Biegler (1988). The control algorithm can be summarized as follows:

Algorithm 5.1

Step 0.	1. Select design parameters $k_{11}, k_{12}, k_{21}, k_{22}$ (or ξ and τ) 2. Select constraint handling parameters k^U_{iC} and k^L_{iC} , $i =1,2,3$ 3. Select optimization weighting factors W_i, $i =1,....,10$
At each sampling time	
Step 1.	Measure outputs y_i (t), $i = 1,2, 3$
Step 2.	Calculate the process inputs u_i (t), $i=1,2,3$ by solving the optimization problem (5.1) subject to Equations 5.4 - 5.14.
Shift to the next sampling time.	

The above optimization problem was applied to heavy oil fractionator and the solution was obtained by using the Optimization toolbox of MATLAB® by the Mathworks, Inc. (1993). The toolbox contains many commands for the optimization of general linear and non-linear functions. The optimization toolbox contains **CONSTR.M** file that finds the constrained minimum of a function of several variables. The program uses sequential quadratic program (SQP) method in which the search direction is the solution of a quadratic-programming problem. From Non-linear Control Design (NCD) toolbox, a non-linear optimization file **NLINOPT.M** was called to run the optimization algorithm which in return calls a routine which converts lower and upper bounds into constraints used by the optimization. An alternative approach to solving non-linear optimization problem is to use the subroutine SOL/NPSOL described by Gill *et al.* (1986) or to use equation-based non-linear optimization software RT-Opt. (real-time optimization) of AspenTech. This later approach was used in Chapter 6 to solve a constrained non-linear multivariable control problem on the reactor-section of a catalytic reforming process.

5.4 Model Parameter Update

In model-based controller, including generic model control (GMC), there is an element of mismatch between the model and the true process. This process-model mismatch leads to deterioration in control performance. There are two types of model mismatch: structure mismatch occurs when the process and the model are of a different nature (e.g., first-order/second-order, or linear/non-linear); parameters mismatch occurs when the numerical values of parameters in the model do not correspond with the true values. Lee *et al.* (1989) presented a process-model mismatch compensation algorithm for model-based control. This algorithm compensated for model errors and updated the model parameters at steady state. Signal and Lee (1992) derived an adaptive algorithm capable of adapting model parameters in a non-linear model. The algorithm was developed within a GMC framework that reduces the effect of larger modelling errors by regularly updating the model parameters.

When the system is minimum phase and in the absence of process model error, the closed loop response by GMC will follow the reference trajectory exactly. However, when the model is inaccurate, the response will deviate from the reference trajectory. The integral term in GMC control law provides a compensation for the process-model mismatch. The control structure need not be changed if the closed loop response by this compensation is satisfactory but this method may not be applicable to all the systems because of the stability and the process constraints. It must be noted that mathematical conditions for accurate tracking and offset free control are much stronger than these specified reference trajectories.

When the mismatch becomes larger, the closed loop response will not be satisfactory. Also for process control point of view and for process industries, it is often required that the closed-loop response exhibit no overshoot and have suitable rise-time. These properties are ensured when a perfect process model is used or the mismatch is small, but these are not guaranteed when the mismatch and any unmeasured disturbances are large. The basic idea behind the development and application of a dynamic parameter update system is to cope with these difficulties and reduce the effect of larger modelling errors by regularly updating the model parameters. In this work, it is assumed that the mismatch is not large and therefore the integral term in GMC control law provides the compensation for the process-model mismatch. The dynamic parameter update system is developed in Chapter 7 for the case of fluid catalytic cracking process.

5.5 Model-predictive Control

5.5.1 Problem Formulation for Linear Control

The model-predictive control algorithm applied here utilizes features from established model-based control algorithms such as dynamic matrix control (DMC) and model algorithm control (MAC) as discussed in detail in Chapter 2. The intent was to use features of each algorithm as they best applied to the problems associated with heavy oil fractionator control. The main characteristics of the control algorithm are summarized here:

1. Discrete step response dynamic models.
2. Reference trajectories to define the desired closed-loop response.
3. Quadratic objective function solved for least squared error.
4. Iterative control calculation for constrained input operation.

These methods are well established and field proven; however, an overall control strategy requires additional programme functions, which are built around the basic MPC algorithm.

Control moves were calculated to minimize a quadratic objective function of the form:

$$\mathbf{J}\,[\Delta u] = \mathbf{E}^T\mathbf{Q}\mathbf{E} + \Delta u\,\mathbf{R}\,\Delta u \qquad\qquad (5.16)$$

where E is the predicted error array, $\mathbf{Q}$ is the control error weighting matrix and $\mathbf{R}$ is the input penalty weighting matrix.

The control move to minimize $\mathbf{J}$ is calculated using least squares

$$\Delta u = (\mathbf{A}^T\mathbf{Q}\mathbf{A} + \mathbf{R})^{-1}\,\mathbf{A}^T\mathbf{Q}\mathbf{E} \qquad\qquad (5.17)$$

where $\mathbf{A}$ is the dynamic matrix composed of step response models. The $\mathbf{Q}$ weighting matrix provides scaling for the control variable errors and allows weighting of one constraint versus another. The **R** weighting matrix penalizes excessive moves and "ringing" of manipulated variable.

The MATLAB® toolbox on MPC was used for the linear control application together with the control techniques mentioned above. These techniques were incorporated in the programme and applied to Shell heavy oil fractionator. The results obtained were compared with the constrained non-linear multivariable control application developed for the same process.

5.5.2 MATLAB®/SIMULINK® Programme for Heavy Oil Fractionator

MATLAB® Programme

In order to run the MATLAB® programme for heavy oil fractionator, enter the following at the MATLAB® commands line:
```
>>load step2
>>mpct
>>dist2
```
Select "start" under the Simulation menu item. 'step2' contains the step response vectors, which will load into workspace. 'mpct' then builds the prediction and control matrices from the step response data. Multivariable controller is the actual dynamic matrix control algorithm
implemented in discrete form. MATLAB® file "mpct.m" contains commands which allow the input of different move suppression factors for manipulated variables. Rest of the procedure follows the standard MATLAB®/SIMULINK® standard commands.

Heavy oil fractionator problem was built in MATLAB®/SIMULINK® software using the features of model-predictive control techniques such as dynamic matrix control. SIMULINK® was used as it is a tool for modelling, analyzing, and simulating a wide variety of physical and mathematical systems, including those with non-linear systems and those which make use of continuous and discrete time. As an extension of MATLAB®, SIMULINK® adds many features specific to dynamic systems while retaining the MATLAB®'s general-purpose functionality. Figure 5.4 shows the SIMULINK® schematic for Shell heavy oil fractionator.

Two disturbances, intermediate reflux duty (d_1) and upper reflux duty (d_2) are shown to enter the column. By clicking twice on these input signals, a window appears which allows the user to input the data for these disturbances. By twice clicking the "column" Figure 5.5 appears on the screen showing the connections of five inputs (manipulated variables) to three outputs (controlled variables) as defined in Figure 5.2 which represents the control problem schematic of heavy oil fractionator.

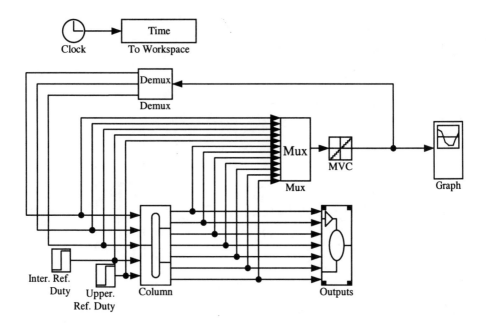

Figure 5.4. SIMULINK® schematic for the Shell heavy oil fractionator

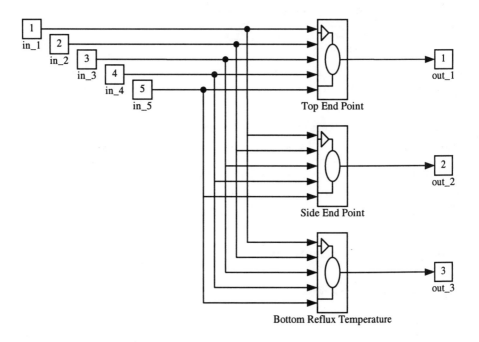

Figure 5.5. Input-output connections for the Shell heavy oil fractionator

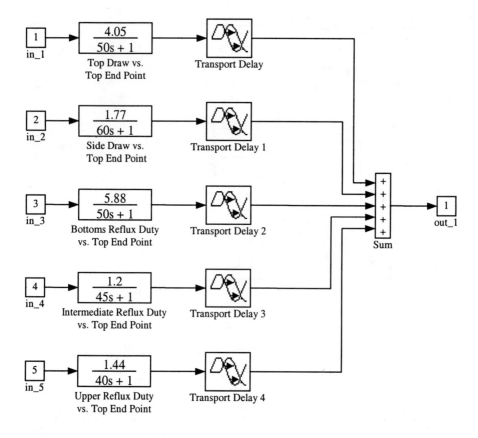

Figure 5.6. SIMULINK® connecting Top End Point to manipulated variables

In Figure 5.5, by clicking twice the controlled variable such as "top end point", another screen appears with Figure 5.6. This figure correlates top end point process model (first-order and dead time) to five manipulated variables as shown in Table 5.1. Similarly clicking twice the other controlled variables in Figure 5.5 connection to other models can be obtained. The model-predictive multivariable control programme for heavy oil fractionator is given in Appendix C.

5.6 Results and Discussions

5.6.1 Simulation Results

In this section the results of simulation studies based upon the four test cases are presented in order to illustrate the performance of the non-linear model-based multivariable control and linear control. In particular the ability of the controller to satisfy the performance criteria such as integral action (*i.e.*, zero offset from setpoint at steady state), satisfaction of process constraints, disturbance rejection, adequate speed of response (*i.e.*, closed loop speed of response between 0.8 and 1.25 of the open loop response), and ability to maintain good control upon loss of feedback measurement. As shown in Figure 5.2, for both controllers: linear and non-linear, the manipulated variables are top draw, side draw and bottom reflux duty. The controlled variables are top and side end points and associated variable is bottom reflux temperature.

The simulation took about four seconds to get a solution for every sample time. The parameters of the non-linear control system are given in Table 5.3. The GMC parameters ξ and τ for the controlled variables were selected from the GMC performance specification curve (Lee and Sullivan, 1988) according to the time constants shown in Table 5.1 and the control objective of keeping the closed-loop speed of response between 0.8 and 1.25 of the open-loop process bandwidth.

Table 5.3. Parameters of the control system

		Non-linear Control System Parameters
ξ	5	for y_1 and y_2
τ (min)	100	for y_1 and y_2
k^U_i	0.5	i= 1, 2, 3
K^L_i	0.5	i= 1, 2, 3
W_i	200	$W_1 = W_2 = W_3 = W_4$
W_i	100	$W_5 = W_6 = W_7 = W_8 = W_9 = W_{10}$

These pair of parameters has explicit effects on the closed-loop response of output variable. The parameter ξ_i determines the shape of the closed-loop response while τ_i determines the speed of the response (large τ_i means slow response). The weighting factors W_i were selected by considering the relative importance of the terms in the objective function. The sampling time was considered 5 minutes.

Control Objectives 1 and 2, defined under Section 5.2.2.1 are analyzed here by introducing arbitrary disturbances of magnitude less than 0.5 in IRD and URD. These disturbances caused top end point to reach to a steady state value 0.32 based on the process model and with disturbances bounded in magnitude by 0.5. Figure 5.7 shows the responses of the controller to positive steps of 0.5 in both the disturbances. The steady state offset specification 0.005 maximum seems to be very stringent and can not be met in the event of analyzer failure. A realistic specification in the event of analyzer failure may be a steady state offset less than and equal to 0.5.

5.6.2 Integrating Top End Point (TEP) Inferential Model

In order to reduce the steady state offset, an inferential model was developed to estimate or predict the top end point values based on the available measurements as shown in Figure 5.8. Using this model, inputs can be calculated which would eliminate end point offset for the nominal model. To determine the effectiveness of this approach, it was assumed that the disturbances can be estimated perfectly at steady state, and steady the offset which results for the worst-case plant. The top

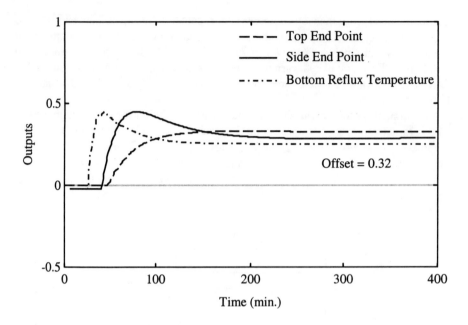

Figure 5.7. Steady-state offset in TEP and SEP for arbitrary disturbances in both the disturbance variables

end point was considered here for the analysis and an inferential model was developed to predict the values in the absence of analyzer.

The top end point prediction is based on the following correlation that can be derived from the steady-state model of the crude distillation. The development issues related to this type of inferential modelling work was discussed in details in Chapter 3. The functional form of correlation is given by:

$$TEP = [\, P, T, \Lambda_1, R/D \,] + \Lambda_2 \tag{5.18}$$

In Equation 5.18, P and T are the column pressure and top tray temperature respectively, R/D is the upper reflux to top draw ratio, Λ_1 is the filter of the first order to reduce the measurement noise and Λ_2 is a bias or trim factor to update the model from the analyzer or laboratory quality results. It is important to note that the Equation 5.18 is non-linear in terms of the upper reflux to top draw ration (distillate ratio). The advantage of using a non-linear model is that it is valid over a wider operating range and reduces the retuning effort. Equation 5.18 does not take into account pressure-temperature compensation. The pressure-temperature compensation was carried out using the non-linear Clausius-Clapeyron equation given in Perry and Green (1984). It was modified in the following form for this particular application:

$$Tcomp = T_k * B_i /(T_k * LnP + B_i) \tag{5.19}$$

where Tcomp is a pressure compensated temperature, T_k is tray temperature in Kelvin and B_i is a Clausius-Clapeyron coefficient. The theory behind the pressure-compensated temperature was discussed in details in Chapter 3. Equation 5.18 can be derived from the steady-state model which receives the scanned values of pressure, temperature, reflux and top draw (distillate) every two minutes from the real-time data acquisition system. Based on these values the model calculates and predicts the quality of the product. The dynamic effects are compensated as the tray temperature at the top of the column and reflux and distillate flows are fast loops. The pressure-compensated temperature obtained from Equation 5.19 compensates for the effect of pressure variation in the column.

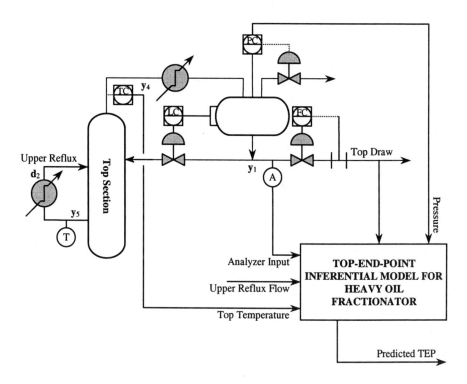

Figure 5.8. Inferential model development procedure of TEP for heavy oil fractionator

The model is updated using laboratory results or the analyzer's input. A system of inferential model update is shown in Figure 5.9, providing laboratory/analyzer result as a basis to update the predicted values and a first-order filter to compensate for the noise in the flow measurements.

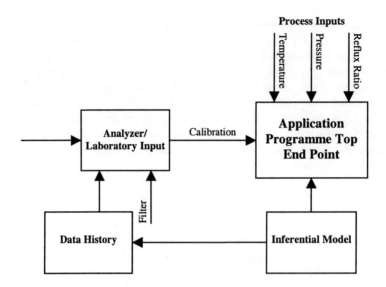

Figure 5.9. System of inferential model update for top end point

In order to determine the feasibility of the above approach, it was assumed that the disturbances can be estimated perfectly at steady-state and studied the offset which results for the worst-case plant. Consider a system of 3x5 matrix of the heavy oil fractionator given in Figure 5.2 and the relation is expressed as:

$$y = Gy_u u + Gy_d d \tag{5.20}$$

where y is the steady-state values, u represents any two of the three manipulated variables, and assuming that d, the disturbances to be known. Gy_u and Gy_d are the nominal models. Then end points, y' predicted by the nominal models, $G'y_u$ and $G'y_d$, are given by:

$$y' = G'y_u u + G'y_d d \tag{5.21}$$

selecting u to make y'=0 implies that

$$G'y_u u = -G'y_d d$$

The top draw and side draw rates are uniquely determined by

$$u = -G'^{-1}y_u u \ G'y_d d \tag{5.22}$$

By substituting Equation 5.22 into Equation 5.20, the corresponding steady-state end point values are given by:

$$y = [Gy_d - Gy_u\, G'^{-1}y_u\, G'y_d]\, d \tag{5.23}$$

If there is no model uncertainty, then $Gy_u = G'y_u$, $Gy_d = G'y_d$ and $y = 0$

This implies that even if the disturbances are known exactly, it is not possible to meet the stated performance requirements because of model uncertainties.

Figure 5.10 shows the controller responses to positive steps of 0.5 in both the disturbances with TEP inferential model incorporated in the system. It can be observed from this figure that the inferential model has reduced the offset from 0.32 to 0.23 and provided a better steady-state performance as compared to analyzer. The minimum steady state offset reported by Campo *et al.* (1990) was 0.41 without incorporating the inferential model.

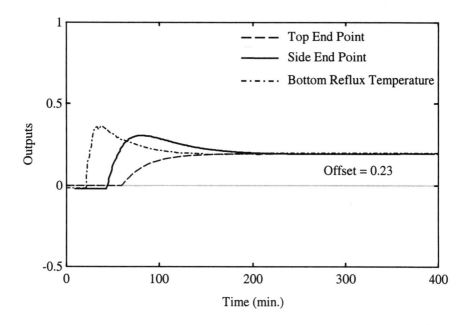

Figure 5.10. Steady-state offset in TEP and SEP for arbitrary disturbances in both the disturbance variables with TEP inferential model

    The estimated values for transfer function model of top draw-TEP (inferential model) are given by K = 4.05, τ = 30 min. and θ =10 min. The value of θ, the deadtime was reduced significantly from 27 minutes to 10 minutes by incorporating the inferential model which provided a better dynamic response in the closed loop quality control system.

    In the following, the linear controller responses for all the four test cases have been compared with the non-linear controller developed in Section 5.3. It can be observed from Test Case 1 (Figures 5.11 and 5.12) that for the nominal plants both controllers exhibit integral action. Settling time for complete disturbance rejection appears to be 200 minutes, which is within the design requirements. The end point responses of both the controllers are qualitatively the same. However, constraints handling performance is better in case of non-linear controller (Figure 5.12) which keeps the top and side end points within the maximum and minimum values of constraints (0.5 and -0.5). Also both controllers tend to trade off the minimization of the bottom reflux duty versus violation of the bottom reflux temperature constraint.

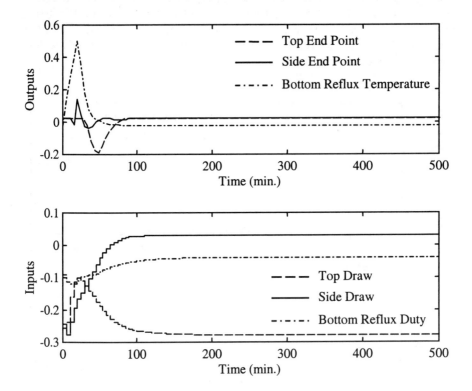

Figure 5.11. Linear controller response to positive steps of 0.5 in both the disturbances (Test Case 1)

The results of Test Case 2 are given in Figures 5.13 and 5.14. In these cases the controller's performance was tested with disturbances of -0.5 in both the intermediate and upper reflux duties. It can be observed from these figures that non-linear controller shows better overall disturbance rejection for the end points. Both the controllers display the ability of optimization of the bottom reflux duty. From these plots it is clear that the two controllers are different in the way in which various performance criteria are traded-off. The non-linear controller, for example, chooses to forgo faster disturbance rejection in favour of lowering the bottom reflux duty (Figure 5.14).

The results of Test Cases 3 and 4 are given in Figures 5.15 to 5.18. In these cases the controller's performance was tested with the step change of 0.5 in the top and side endpoints respectively. The end point responses of both the controllers are similar. Very little manipulated variable action is observed to accomplish disturbance rejection.

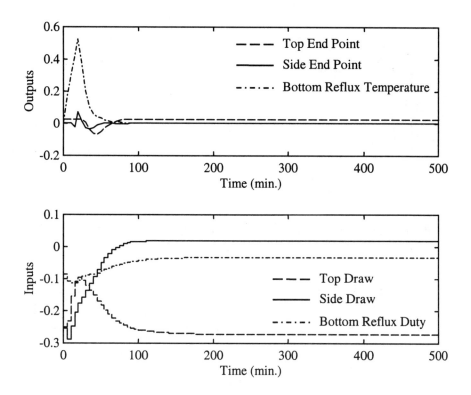

Figure 5.12. Non-linear controller response to positive steps of 0.5 in both the disturbances (Test Case 1)

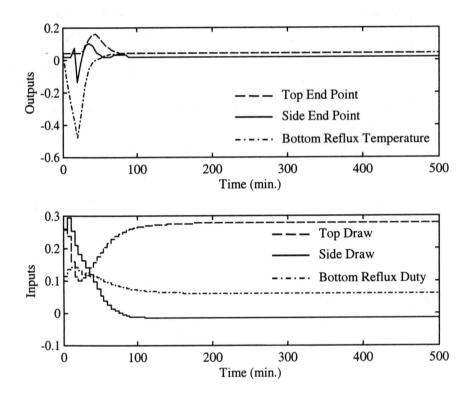

Figure 5.13. Linear controller response to negative steps of -0.5 in both the disturbances (Test Case 2)

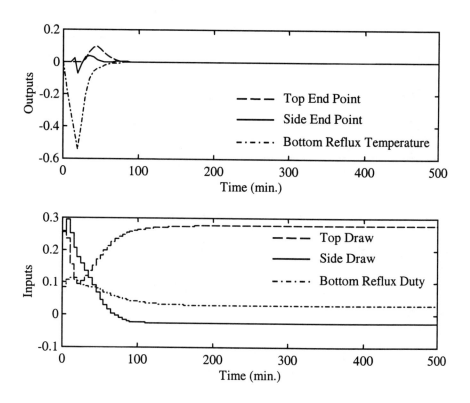

Figure 5.14. Non-linear controller response to negative steps of -0.5 in both the disturbances (Test Case 2)

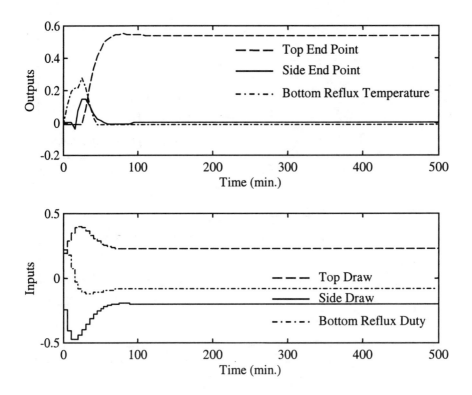

Figure 5.15. Linear controller response to step change of 0.5 in top end point (Test Case 3)

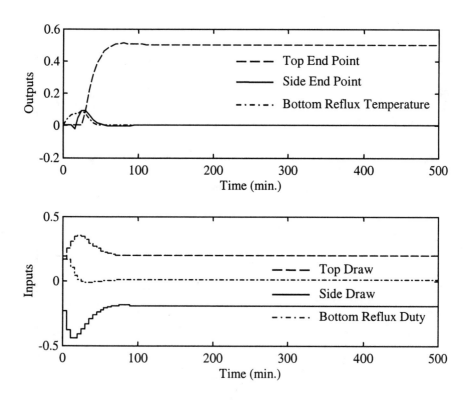

Figure 5.16. Non-linear controller response to step change of 0.5 in top end point (Test Case 3)

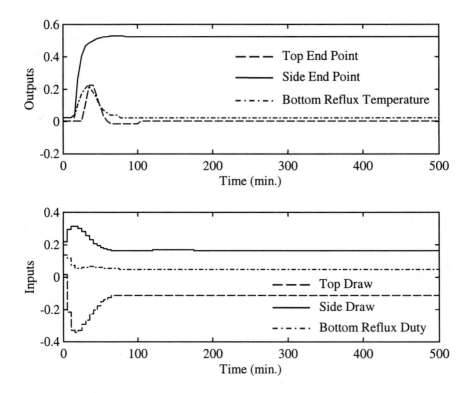

Figure 5.17. Linear controller response to step change of 0.5 in side end point (Test Case 4)

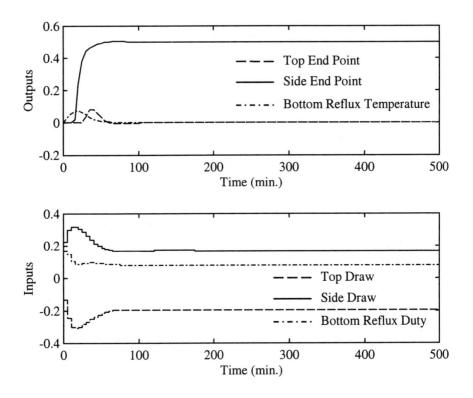

Figure 5.18. Non-linear controller response to step change of 0.5 in side end point (Test Case 4)

5.6.3 Real-time Implementation Results

The non-linear control strategies were tested on a 120,000-barrels/day crude distillation unit. Since the linear controller's results were available on this unit, these were used to compare the non-linear controller's performance. The comparison was limited to two main controlled variables defined in Shell heavy oil fractionator's control problem. Those variables were top end point (naphtha final boiling point) and side end point (kerosene flash point). The same parameters of Table 5.3 were used for real-time application.

The desired trajectory of the process outputs were determined using the tuning rules by Lee and Sullivan (1988). The control application was first turned on and tested in an open-loop mode (*i.e.*, the outputs were suppressed). Open loop testing includes verifying process inputs and checking controller move sizes and direction. Next, the controls were tested in closed-loop mode using tuning constants gathered from the simulation model and step response test. Closed-loop testing was performed within a narrow operating range until confidence was gained. It was also noted that the form of optimization problem is well structured since a slack variable is added to each control law equation to ensure that a solution to the set of equations does in fact exits. It is important to note that if the control is implemented at a reasonable frequency, the solution of the NLP is very fast (3 to 4 iterations) since the current control settings and slack variables provide a good initial estimate of the solution vector. Further, when the control system fails or turns itself off, the basic control system continues to function.

As discussed in Section 5.3, two types of tuning parameters were used; the K^U_{1C} and K^L_{1C}, specifying the maximum speed of approach towards the constraint bounds and weighting factors w, reflecting the relative importance of the outputs and the constraints. The tuning parameters K^U_{1C} and K^L_{1C} can be compared with the move suppression factor, a term used in linear multivariable controller design such as dynamic matrix control (DMC). Changing the move suppression factor on a manipulated variable causes the controller to change the balance between movement of that manipulated variable, and error in the dependent variables.

The role of weighting factors W, is similar to the equal concern factor in linear multivariable control framework such as (DMC). In tuning analysis, it is used to set the priority on controlled variables. In this application for example, the side end point (keroFlash point) was given higher priority over the top end point (naphtha final boiling point).

Figure 5.19 shows the performance of the two controllers in controlling the top end point (naphtha final boiling point) quality. The top end point was required to control between the limits (150 –155)°C. The non-linear controller shows a better performance in keeping the top end point between the lower and upper constraints. It also demonstrates better overall disturbance rejection for the top end point. The real-time results are in agreement with the simulation results shown in Figures 5.11 to 5.14.

Figure 5.20 shows the performance of linear and non-linear controllers in controlling the side end point (kerosene flash point) at various modes of operation for kerosene flash points (65°C, 55°C and 40°C). This figure also shows that both controllers performed well to keep the kerosene flash point at 65°C, however, linear controller did not perform well when the mode of operation (kerosene flash point at 40°C) was changed. It may be due to the fact that the process identification test performed at one set of operating conditions is not valid on other set of conditions as the process dynamics changed and the process gains are different.

The performance of non-linear controller was far better when the mode of operation was changed. This also demonstrates the ability of non-linear controller to control the product specification on targets for a wide range of operating conditions.

5.7 Conclusions

The heavy oil fractionator control problem proposed by Prett and Morari (1987) was solved in this chapter. The solution to the control problem considered the dynamic process model identified by open-loop testing of the plant and constructed a constrained non-linear optimization problem as the bottom reflux duty was required to be minimized.

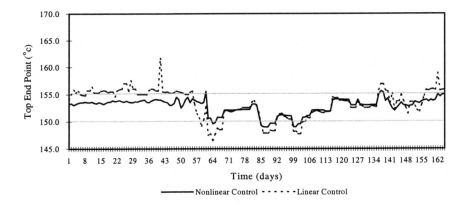

Figure 5.19. Performance of linear and non-linear controllers in controlling the top end point (naphtha final boiling point) of crude distillation

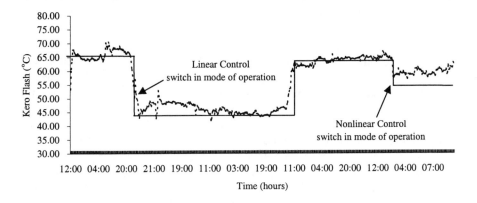

Figure 5.20. Performance of linear and non-linear controllers in controlling the side end point (kerosene flash point) of crude distillation

The control problem of the heavy oil fractionator was analyzed with an extension of the conventional model-predictive control (MPC) algorithm using MATLAB® control features. The simulation results of the linear MPC were compared with the non-linear model-based multivariable control. It has been shown that non-linear multivariable control has provided better overall disturbance rejection for the end points. It has selected to forgo faster disturbance rejection in favour of lowering the bottom reflux duty. It has also demonstrated that by incorporating the top end point inferential model in control strategies, the system has reduced the offset and provided a better steady state performance in case of analyzer failure. The simulation studies conducted show that both controllers exhibit good qualitative robust performance characteristics. However, non-linear controller, due to its flexibility in dealing with constraints is well suited for solving this problem.

In real-time, the non-linear control strategies were tested on a 120,000 barrels per stream day (BPSD) crude-fractionator. The performance of the non-linear controller was compared with the linear controller for two important cases: controlling the top end point quality (naphtha end point) and controlling the side end point (kerosene flash point) at different mode of operation. It has been demonstrated by the real-time results in Figures 5.19 and 5.20 that the non-linear controllers have a better ability to keep the end points targets within the lower and upper constraints and to control the product qualities to specification for a wide range of operating conditions.

The main contribution of this work is to incorporate dynamic process model obtained from the open-loop tests in non-linear constrained optimization strategies and to apply it to heavy oil fractionator. In addition, an inferential model for top end point (TEP) was developed and integrated with process model showing a better response in case of analyzer failure. The MATLAB® toolbox on model-predictive control (MPC) was used for the linear control application incorporating the control techniques in the program such as control variable compensation and prediction trend correction. The presence of hard constraints in the optimization problem makes the overall system non-linear even though the process dynamics are assumed linear.

For the non-linear system, constrained non-linear optimization strategies were applied in real-time to 120,000 BPSD crude-fractionator and the solution was obtained by using the Optimization toolbox of MATLAB®. The **CONSTR.M** file was used to find the constrained minimum of the objective function by using the sequential quadratic program (SQP) method. From Non-linear Control Design (NCD) toolbox, a non-linear optimization file **NLINOPT.M** was used to run the optimization algorithm. This approach, however, points to the deficiencies of optimizing multiple performance criteria via a lumped objective function approach. A more appropriate way for performing multiobjective optimization would be to use a framework in which different objective criteria are handled at different levels. This approach will be recommended in Chapter 8 as a future direction of research.

CHAPTER 6

CONSTRAINED NON-LINEAR MULTIVARIABLE CONTROL OF A CATALYTIC REFORMING PROCESS

The ability to handle constraints, which generally are non-linear functions of input, state and output variables, is an important feature for industrial control algorithms when they are applied to real processes. In this chapter, the basic assumptions and concepts about the non-linear process and constraints are defined. A non-linear constrained optimization strategy for handling the constraints is proposed and applied to the catalytic reforming reactor section which minimizes a function of the slack variables and allows a balance to be sought between the process outputs tracking their reference trajectories and constraint violation. In this manner, constraint variables can be controlled to the constraint limits as well as to the rate of approach to the constraint limit. Slack variables defining the trajectory departure from the chosen specification curves are added to the generic model control (GMC) control law for both the control variables and the constraint variables. The solution of the problem then becomes a non-linear constrained optimization, which minimizes a function of the slack variables.

A dynamic model of the catalytic reforming process was developed by assembling the mass and energy balances on the system of reactions and then used for non-linear multivariable control application on the reactor section to provide target values for the reactor inlet temperature. The model results were tested in real-time application and the results were used to improve the dynamic response of the model. The constrained non-linear multivariable controller controls the weighted average inlet temperature (WAIT) while respecting the heater tube temperature constraints. WAIT is set by an octane inferential model forming a closed-loop WAIT/octane control.

6.1 Introduction

Many practical control problems possess constraints on the input, state and output variables. Although model-based process control has drawn considerable attention in process control because of its good performance characteristics, none of the techniques were originally designed with explicit constraint handling methods. The ability to handle constraints is essential for any algorithm to be implemented on real processes. Thus, strategies for constraint handling within model-based algorithms has become one of the more popular research topics.

The model-predictive control algorithms, represented by dynamic matrix control (Cutler and Ramaker, 1980) and model algorithm control (Rouhani and Mehra, 1982) were based on a discrete convolution model of the process and are basically linear model-based algorithms. Their control law without constraints was constructed as a typical optimization problem, so naturally when the constraints were included, the control law was constructed as a typical linearly constrained optimization problem, for which any of the methods for linearly constrained optimization can be used. Therefore, there are still no constrained algorithms available which use the non-linear process model and general non-linear constraints directly without a linearization step. Both linear programming (Chang and Seborg, 1983) and quadratic programming (Little and Edgar, 1986) have been applied in model-predictive control for processes with linear constraints. Internal model control (Garcia and Morari, 1982, 1985), as a control framework, is also linear model-based control. Linear programming (Brosilow et al., 1984) and quadratic programming (Ricker, 1985) were also applied for constraint treatment in a similar manner to that applied for the model-predictive control algorithms. Economou et al. (1986) extended IMC to non-linear lumped parameter systems by an operator approach. The key issue was inversion of the non-linear process model in the control law, and a Newton-type method was adopted. Li and Biegler (1988) have recently extended this operator approach to deal with linear input and state variable constraints by a successive quadratic programming (SQP) strategy.

In a different approach, a control framework for both linear and non-linear systems, Generic Model Control (GMC) has been developed in the time domain (Lee and Sullivan, 1988). The control law employs a non-linear process model directly within the controller (Lee, 1991). Also an, integral feedback term is included such that the closed loop response exhibits zero offset. Similar approaches have been proposed by Liu (1967), Balchen et al. (1988), and Bartusiak et al. (1989). For each of the output variables, there are two controller performance parameters which specify the shape of the closed loop system response. These parameters are selected a priori by considering the open-loop characteristics such as the process time constant and process deadtime together with the sampling time interval. GMC has been extended to compensate for process/model mismatch (Lee et. al. 1989a) and deadtime compensation (Lee et al., 1990), but the strategies for constraint handling within the GMC framework have not yet been explored extensively and have not been applied to complex processes in petroleum refinery. Brown et. al. (1990) has applied these strategies on an ideal Stirred Tank Reactor (STR) with reversible exothermic reaction.

The neural-network-model-predictive control (NNMPC) is another typical and straightforward application to non-linear control. Ishida and Zhan (1995) proposed a neural model-predictive control (NMPC) method for the one-step predictive control of MIMO processes, which may be classified as an IMC technique, based on the steepest descent optimization algorithm. Mills et. al. (1995) also discussed similar approaches for adaptive model-based control using neural networks. Spangler (1994) reported that non-linear optimization codes (e.g., SQP) can be integrated with a virtual analyzer neural network model to compute setpoints for process inputs that improve the values of the sensed quantities (reduce emission) while satisfying process performance and/or cost constraints. Recently Keeler *et al.* (1996) has developed "The Process Perfecter" software based on neural network modelling techniques, which enables closed-loop control and optimization of non-linear processes.

The basic idea behind the development of Process Perfecter is the same as GMC as it also combines non-linear multivariable control with non-linear constrained optimization models, it can be used in non-linear processes where traditional linear control is not applicable.

This chapter provides details of implementation of constraint non-linear multivariable control and solves a non-linear optimization problem on a semi regenerative catalytic reforming reactor section. The main contributions of this work are to integrate the octane inferential model with non-linear multivariable controller; and develop a non-linear constrained optimization algorithm, test and implement the algorithm in real-time on a highly non-linear catalytic reforming process. The main objective of this real-time application to catalytic reforming reactor section is to control WAIT/octane at target values, and for a given octane and feed rate, maximize the yield of reformate consistent with the key operating constraints.

6.2 Process Constraints Classifications

Process constraints can be imposed on the input, state, and output variables or a combination thereof in order to represent physical limitations of the equipment, economic considerations, safety and environmental regulations, product specifications or even human preferences (Prett and Garcia, 1987). There are many different methods to clasify constraints. One possible method is to place them into three different categories.

1. Input Constraints:

$$u_L \leq u \leq u_U \tag{6.1}$$

These types of constraints are usually caused by the physical limitations of the equipment.

2. Input Movement Constraints:

$$\Delta u_L \leq u(t +\Delta t) - u (t) \leq \Delta u_U \tag{6.2}$$

This type of constraint is usually caused by the physical limitations of the equipment, but it is associated with the dynamic properties of the equipment and dependent on the value of the sampling time interval Δt.

3. Functional Constraints:

$$C_L \leq C(y, x, u, t) \leq C_U \tag{6.3}$$

where C, C_L and C_U are vectors of dimension p, each element of C is generally a non-linear function of y, x, u and t. These constraints are usually derived from economic and process considerations.

Using other classifications, constraints can also be derived according to their properties (Prett and Garcia, 1987) as follows:

1. Hard Constraints: No dynamic violations of the bounds are allowed at any time.

2. Soft Constraints: Violations of the bounds are temporarily permitted in order to satisfy other heavily weighted criteria.

The input and input movement constraints defined above are hard constraints, while most of the functional constraints are soft constraints. In modelling real processes, there will exist a certain degree of process/model mismatch due to the assumptions made in deriving the process model and the effects of the unmodelled process disturbances. Therefore it is recommended that the constraints involving the process state and output variables be treated as soft constraints.

For GMC, handling the input and input movement constraints is straightforward. These constraints simply limit the control action determined by GMC control law, resulting in a slower output response than expected. The following section will present an optimization approach, which gives a unified treatment of the functional constraints as well as the input and input movement constraints.

For the convenience of the development to follow in Section 6.3, some definitions are introduced below:

Definition 1: System Status. S = (**y**, **x**, **u**, t) is called a system status at time t, if **y**, **x**, **u** are process output, state and input variables at time t respectively

$S^\circ = (y^\circ, x^\circ, u^\circ, t^\circ)$ is called the initial system status of the process, where t° is the initial time and y°, x°, u° are the initial process output, state and input variables.

Similarly, S* = (**y***, **x***, **u***, t*) is called the final system status, where **x*** and **u*** are the process state and input variables when the process output has reached the setpoint **y***. It is to be observed that even after y has settled at its setpoint **y***, **x*** and **u*** can still be a function of time for rejecting disturbances.

Definition 2: Feasible Region. Given the functional constraints as described in Inequality 6.3, the feasible region Ω is defined as the set of all system status (**y**, **x**, **u**, t) which satisfies Inequality 6.3:

$$\Omega = \{ SIS = (y, x, u, t); C_L \leq C(y, x, u, t) \leq C_U \} \tag{6.4}$$

In Section 6.3, it is assumed that the final system status S* is inside the feasible region Ω, but the initial system status S° may be outside it.

6.3 Constraint Non-linear Multivariable Control

6.3.1 Control Theory and Design

It was mentioned in Section 6.2 that the GMC control law can be constructed as an optimization problem. Thus a constrained optimization structure can be developed for GMC when process constraints are present. In this section, a method is proposed where slack variables defining the variable departure from the chosen GMC specification curves are added to the GMC control law for both the controlled variables and constraint variables. Selecting the weighting factors on these slack variables and defining a control objective function, which is dependent on these weighted slack variables, allows the controller to achieve the desired compromise between constraint violation and setpoint tracking. The solution to the problem becomes a non-linear constrained optimization that does not rely on a multi-time step model prediction. The main advantage of the proposed approach is that a single time step control law results in a much smaller dimensional non-linear program than previous methods do. In the following, an optimization problem is constructed which incorporates the process constraints.

Consider the case where the number of outputs and the number of inputs are equal and the non-linear process is defined by the following equations:

$$\frac{d\mathbf{x}}{dt} = \mathbf{f}(\mathbf{x,\ u,\ d},\ t) \qquad (6.5)$$

$$\mathbf{y} = \mathbf{g(x)} \qquad (6.6)$$

where $\mathbf{x}$ is the state vector of dimension n, $\mathbf{u}$ is the input vector of dimension m, $\mathbf{d}$ the disturbance vector of dimension 1, and $\mathbf{y}$ is the output vector of dimension m. In general $\mathbf{f}$ and $\mathbf{g}$ are vectors of non-linear functions. Equation 6.5 is the general form of the model of a non-linear process.

In GMC, the rate of change of the controlled variables is set equal to the proportional and integral term (Lee and Sullivan, 1988).

$$dx/dt = \mathbf{k}_1(\mathbf{y}_{sp} - \mathbf{y}) + \mathbf{k}_2 \int (\mathbf{y}_{sp} - \mathbf{y})dt \qquad (6.7)$$

where k_1 and k_2 are diagonal matrices of GMC parameters. Substituting Equation 6.5 into Equation 6.7 yields the GMC control law:

$$f(\mathbf{x},\mathbf{u},\mathbf{d},t) = \mathbf{k}_1(\mathbf{y}_{sp} - \mathbf{y}) + \mathbf{k}_2 \int (\mathbf{y}_{sp} - \mathbf{y})dt \tag{6.8}$$

It is desirable to have the system operate within the feasible region such that for the p constraints:

$$\mathbf{C}_L \leq \mathbf{C}(\mathbf{y}, \mathbf{x}, \mathbf{u}, t) \leq \mathbf{C}_U \tag{6.9}$$

where $\mathbf{C}$, $\mathbf{C}_L$ and $\mathbf{C}_U$ are vectors of dimension p, each element of $\mathbf{C}$ is generally a non-linear function of $\mathbf{y}$, $\mathbf{x}$, $\mathbf{u}$ and t. These constraints are usually derived from economic and process considerations. The $\mathbf{C}_L$ and $\mathbf{C}_U$ are the lower and upper bound of vector $\mathbf{C}$.

Both input constraints and input movement constraints are also defined for the n controls, such that:

$$\mathbf{u}_L \leq \mathbf{u} \leq \mathbf{u}_U \tag{6.10}$$

where $\mathbf{u}$ is an input vector and $\mathbf{u}_L$ and $\mathbf{u}_U$ are the lower and upper bounds of $\mathbf{u}$.

$$\Delta\mathbf{u}_L \leq \mathbf{u}(t + \Delta t) - \mathbf{u}(t) \leq \Delta\mathbf{u}_U \tag{6.11}$$

where $\Delta\mathbf{u}_L$ and $\Delta\mathbf{u}_U$ are the lower and upper bounds on input movement vector $\Delta\mathbf{u}$. If the state of the process is such that it lies outside the feasible region or that the current state variable trajectories will violate the given process constraints, then it is desirable to operate the system such that the rate of change of the constraint variable $\mathbf{C}$ approaches its constraint value according to the GMC reference trajectory:

$$\frac{d\mathbf{C}}{dt} = \mathbf{K}_{1C}(\mathbf{C}_U - \mathbf{C}) \tag{6.12}$$

and

$$\frac{d\mathbf{C}}{dt} = \mathbf{K}_{2C}(\mathbf{C} - \mathbf{C}_L) \tag{6.13}$$

where $\mathbf{K}_{1C}$ and $\mathbf{K}_{2C}$ are p x p diagonal matrices, that specify the maximum speed of approach towards the constraints bounds. The p dimensional slack variables vectors, λ^-_c and λ^+_c, defining the variables departure from the chosen specification curves are added to Equations 6.12 and 6.13 for the constraint variables. The constraint variables should be normalized to ensure proper scaling of the problem and allow constraints on different variables to be handled in the same objective function. For the case of the known constraint dynamics, Equations 6.12 and 6.13 become:

$$\frac{d\mathbf{C}}{dt} - \lambda_c^- \le \mathbf{K}_{1C} (\mathbf{C}_U - \mathbf{C}) \qquad \text{and} \qquad (6.14)$$

$$\frac{d\mathbf{C}}{dt} + \lambda_c^+ \le \mathbf{K}_{2C} (\mathbf{C} - \mathbf{C}_L) \qquad (6.15)$$

where λ_c^- and λ_c^+ represent the variables departure from the chosen specification curves for the upper and lower constraints respectively. These slack variables, values of which will be chosen in the proposed optimization problem to be developed, allow the control algorithm to "trade-off" performance versus constraint violation. Defining a GMC control law for the constraint variables not only allows the system to ensure that the constraint is not violated but provides control over the rate of approach to the constraint through the selection of $\mathbf{K}_{1C}$ and $\mathbf{K}_{2C}$.

The inclusion of the constraint on $d\mathbf{C}/dt$, as given by Equations 6.14 and 6.15, rather than the constraint, $\mathbf{C}$, itself, in the non-linear program (NLP) formulation leads to significant computational savings. Previously a multi-time step prediction of the controlled response was required in order to ensure that no constraints are violated in the future. This leads to a large dimensional NLP formulation. By constraining the rate of approach to a constraint limit, the NLP formulation is essentially reduced to a single time step problem.

A set of performance slack variables can also be incorporated into the GMC performance curves to denote the systems efficiency in terms of setpoint tracking. If the two m dimensional performance slack variables vectors, λ_p^- and λ_p^+, are defined to express the systems negative offset and positive offset from the pre-specified response trajectory, the GMC control law for the system performance can be written as:

$$\frac{\partial \mathbf{g}}{\partial \mathbf{x}} \mathbf{f}(\mathbf{x}, \mathbf{u}, \mathbf{d}, t) + \lambda_p^+ - \lambda_p^- = \mathbf{k}_1 (\mathbf{y}_{sp} - \mathbf{y}) + \mathbf{k}_2 \int_{t_0}^{t} (\mathbf{y}_{sp} - \mathbf{y}) dt$$

$$(6.16)$$

where y_{sp} represents the process setpoints. The constrained multivariable control problem can be solved as a non-linear constrained optimization problem, which minimizes a function of the slack variables. The overall GMC control problem can now be formulated as the following single step non-linear optimization problem:

NLP1:

Choose: $\mathbf{u}, \lambda_p^+, \lambda_p^-, \lambda_c^+, \lambda_c^-$

To minimize:

$$J = \mathbf{W}^-_p\, \lambda^-_p + \mathbf{W}^+_p\, \lambda^+_p + \mathbf{W}^-_c\, \lambda^-_c + \mathbf{W}^+_c\, \lambda^+_c \qquad (6.17)$$

where $\mathbf{W}$ are weighting matrices with elements such that

$w_{ii} \geq 0,\ w_{ij} = 0,\ i \neq j.$

Subject to:

$$\frac{\partial \mathbf{g}}{\partial \mathbf{x}}\, \mathbf{f}(\mathbf{x}, \mathbf{u}, \mathbf{d}, t) + \lambda^+_p - \lambda^-_p = \mathbf{k}_1\, (\mathbf{y}_{sp} - \mathbf{y}) + \mathbf{k}_2 \int_{t_0}^{t} (\mathbf{y}_{sp} - \mathbf{y})dt$$

$$(6.18)$$

$$\frac{d\mathbf{C}}{dt} - \lambda^-_c \leq \mathbf{K}_{1C}\, (\mathbf{C}_U - \mathbf{C}) \qquad (6.19)$$

$$\frac{d\mathbf{C}}{dt} + \lambda^+_c \leq \mathbf{K}_{2C}\, (\mathbf{C} - \mathbf{C}_L) \qquad (6.20)$$

$$\mathbf{u}_L \leq \mathbf{u} \leq \mathbf{u}_U \qquad (6.21)$$

$$\Delta \mathbf{u}_L \leq \mathbf{u}(t + \Delta t) - \mathbf{u}(t) \leq \Delta \mathbf{u}_U \qquad (6.22)$$

$$\lambda^+_p \geq 0,\ \lambda^-_p \geq 0, \lambda^+_c \geq 0, \lambda^-_c \geq 0 \qquad (6.23)$$

The overall problem described above can be solved as a single time step NLP. The Equation 6.23 shows the slack variables for the trajectories of the outputs and the trajectories of the constraints. From Equations 6.18 and 6.23, it is clear that positive values of λ^-_p or λ^+_p represent the difference between output response and the GMC reference trajectory. This difference is a measurement of the control performance degradation. Positiveλ^-_c or λ^+_c, in Equations 6.19 and 6.20, is the difference between the actual rates of changes of the constraints beyond the specified maximum rate of approach towards the constraint bounds. The larger they are, the more likely the constraints will exceed their bounds in the future. Thus they are the measurement of the potential constraint violation. All of these λ's are desired to be minimized.

The slack variable weighting factors, $\mathbf{W}$, reflect the relative importance of the outputs and constraints. Thus, if the constraint control is more important than the quality control, then:

$$\mathbf{W}^+_c,\ \mathbf{W}^-_c \gg \mathbf{W}^+_p,\ \mathbf{W}^-_p \qquad (6.24)$$

Conversely, if the quality control is more important than the constraint control, then:

$$\mathbf{W^+}_p, \mathbf{W^-}_p \gg \mathbf{W^+}_c, \mathbf{W^-}_c \qquad (6.25)$$

for which the functional constraints virtually have no effect on the closed-loop responses of the system.

An alternative approach to solve dC/dt is based on the assumption that the constraints will move to their constraint value in a first-order manner. A simple estimate of the time response of the constraint variables in moving from one steady state to another can be given as:

$$\frac{d\mathbf{C}}{dt} \cong (\mathbf{C}^{AIM} - \mathbf{C}) \times \frac{1}{\tau_c} \qquad (6.26)$$

$$\mathbf{C}^{AIM} = \mathbf{C}(\mathbf{y}^{AIM}, \mathbf{x}^{AIM}, \mathbf{u}, t) \qquad (6.27)$$

$$\mathbf{f}(\mathbf{x}^{AIM}, \mathbf{u}, \mathbf{d}, t) = 0 \qquad (6.28)$$

$$\mathbf{y}^{AIM} = \mathbf{g}(\mathbf{x}^{AIM}) \qquad (6.29)$$

where τ_c is the estimated time constant of the constraint. C^{AIM} is the steady state value of the constraint corresponding to the present control action. This approach has been used successfully in the unconstrained problem by Cott et al. (1989).

The fact that separate GMC control specification curves have been defined for the constraints as well as for the controlled variables provides a great deal of flexibility in the controller design. The weighting terms are responsible for assigning a priority level to the various control objectives, while the GMC specification parameters are able to define the trajectories of the constraint paths. Therefore, if the setpoint of a given control variable is increased above a heavily weighted upper variable constraint, then the control variable trajectory will deviate from its pre-defined GMC specification curve to follow the GMC trajectory defined for the constraint variable. It is the degree of flexibility in establishing the proper balance between the violation of the constraint variables and the deterioration of the control performance coupled with the ability to predefine the response trajectories for both the controls and constraints which are the merit of this technique.

The optimization problem, which arises, can be solved using a non-linear constrained optimization problem algorithm. The form of optimization problem is well structured since a slack variable is added to each control law equation to ensure that a solution to the set of equations does exist. If the control is implemented at a reasonable frequency, the solution of the NLP is very fast since the current control settings and slack variables provide a good initial estimate of the solution vector.

From the above proposals, an algorithm is summarized as follows:

Algorithm 6.1

Step 1.	Select values of ξ and τ for GMC controller and W^+_p, W^-_p, W^+_c, W^-_c, K_{1C}, K_{2C}.
Step 2.	Using current values of x, u, d, y, t evaluate $\partial g/\partial x$, f, C, dC/dt. Substitute these values into Equations 6.17 to 6.23.
Step 3.	Determine the present control action by solving the optimization problem (6.17) subject to Equations 6.18 to 6.23.
Step 4.	Implement the control action u.
Step 5.	Return to Step 2 at the next sampling time.

Algorithm 6.1 involves the use of a constrained non-linear programming method, but it is applicable to non-square systems and those systems where GMC control law has no solution and the control law must be constructed as shown in Section 6.3.1 in details.

6.3.2 Selection of Design Parameters

6.3.2.1 Selection of K_{1C} and K_{2C}

K_{1C} and K_{2C} specify the maximum speed of approach towards the constraints bounds. Larger values allow the constraints to approach the bounds faster and therefore with more chance of the bounds being exceeded. If K_{1C} and K_{2C} are smaller, the control action will be suppressed more which gives the controller more chance to avoid the future constraint violation, but results in slower dynamics of the closed-loop system. In practice, for every process K_{1C} and K_{2C} should be tuned such that the constraint violation is avoided with appropriate dynamics of the closed-loop system. The values of K_{1C} and K_{2C} are chosen *a priori* using the GMC specification curves.

6.3.2.2 Selection of W

As discussed in the Section 6.3, the slack variable weighting factors, W, of Equation 6.17 reflect the relative importance of the outputs, the constraints, and the cost of making changes in the manipulated variables.

A balance between the performance degradation of different outputs (λ^-_p or λ^+_p) can be made by assigning different values for the elements of W^+_p and W^-_p, while balance between the potential violation of different constraints (λ^-_c or λ^+_c) can be made by assigning different values for the elements of W^+_c and W^-_c. The different scales of the outputs and constraints should be considered.

Trade-off between the performance degradation of the outputs and the potential violation of the constraints can be made by assigning different magnitudes on W_p (W^+_p and W^-_p) and W_c (W^+_c and W^-_c). For hard constraints, W should be selected by Equation 6.24. $W_c = 0$ is equivalent to ignoring the constraints completely. Comparable magnitudes of W_p and W_c offer a wide range of possibilities on the trade-off between control performance and constraint violation.

6.4 Application to Catalytic Reforming Process

6.4.1 Dynamic Model of the Process

6.4.1.1 Introduction

The catalytic reforming process is used to increase low-octane naphtha fractions to high-octane reformate for use as a premium motor fuel blending component or as a source of aromatic hydrocarbons. Although a number of reactions take place during reforming, the predominant is the dehydrogenation of naphthenes to form aromatics. Some of these aromatics are isolated to become petro-chemical feedstocks, but most become motor fuel blending stock of high antiknock quality.

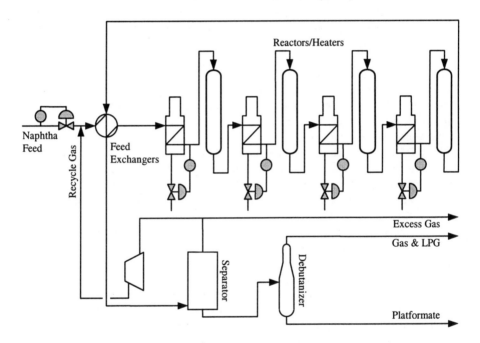

Figure 6.1. Schematic of catalytic reforming process

Most feedstocks for reforming are hydrotreated first to remove arsenic, sulphur and nitrogen compounds. Otherwise these compounds would poison the reforming catalyst. The cost of the hydrotreating step is easily justified by extending the life of the reforming catalyst. Hydrogen is a by-product of catalytic reforming. Some of this hydrogen is recycled to sustain reformer reactor pressure and to suppress coke formation (Heck, 1980). A schematic of the catalytic reforming process is shown in Figure 6.1. Feed to the reactors is combined with recycle hydrogen and preheated with reactor effluent. The feed is then brought up to reaction temperature in a direct-fired heater before entering the first of four reactors arranged in series. Since the net reforming reactions are endothermic, additional heat must be provided to the effluent from one reactor before entering the next. This is accomplished by direct-fired heaters on the inlet to the subsequent three reactors. Temperature controls on the outlet of each of the heaters provide the basis for controlling heat input to each of the reactors.

Modelling and simulation work on the catalytic reforming process is carried out using Smith's kinetic scheme (Smith, 1970) and the model results were tested on an existing plant. The results from the dynamic model were reviewed to improve the processing conditions and the same model was used within the catalytic reforming multivariable control applications.

Smith's kinetic scheme was used to develop a dynamic model for multivariable control applications because it is simple to develop a dynamic model for control purposes from that scheme as compared to other complex kinetic schemes available in literature. The simple kinetic scheme provides a short-cut approach for estimating the product yield of the catalytic reforming process.

6.4.1.2 Model Development
There are a number of different methods for modelling the catalytic reforming process and the main idea behind all these models is to determine the operating conditions of the reforming unit and to predict the yield of reformate, and temperature profile accurately.

A model must predict behaviour not only within the reactor, but in the auxiliary areas of the unit as well. It should also consider the complex nature of the process and the reactions which take place during the process of reforming.

Other kinetic schemes available in literature are very complex as these schemes consider secondary reactions taking place in catalytic reforming (Ansari, 1981). From fundamental point of view kinetics contribute to a better understanding of the reaction mechanisms and of the effect of catalyst upon these. From a practical point of view accurate kinetic equations are of crucial importance for the reliable design and simulation of reactors. Kinetic analysis of reforming process is still following very complex lines (Weekman Jr., 1987). Multicomponent reactions accompanying catalytic reforming provide an insight into the more complex relations of a commercial operation but the component data for specific rate constant, pressure exponent and equilibrium constant for several of reforming reactions is not available in literature.

A mathematical model was developed by assembling the mass and energy balances on the system of reactions. The mass balance provides the variation of concentration of the components selected along the reactors and the energy balance gives the variation of temperature. Mass and energy balances were carried out on one element of radial section of the reactor and then integrated over the whole reactor. Calculation of a reaction rate for all of the feed components over several reactors with declining temperature is the heart of establishing a model. The following is the kinetic scheme proposed by Smith (1970) for the process of catalytic reforming.

$$\frac{n}{3} \text{ Gas} \xleftarrow{\quad k_5 \quad} \text{Naphthene} + H_2 \xleftrightarrow{\quad k_1/k_2 \quad} \text{Aromatics} + 3H_2$$

$$k_3 \uparrow \downarrow k_4 \qquad\qquad\qquad (6.30)$$

$$\frac{n}{3} \text{ Gas} \xleftarrow{\quad k_6 \quad} \text{Paraffin}$$

Taking material balances on all the four components and energy balances on a differential element dw, of catalyst, the following system of differential equations was obtained:

$$\frac{dy_N}{dw} = \frac{MB}{GB}(-r_1 + r_2 - r_4 + r_3 - r_5) \qquad\qquad (6.31)$$

$$\frac{dy_P}{dw} = \frac{MB}{GB}(-r_3 + r_4 - r_6) \qquad\qquad (6.32)$$

$$\frac{dy_A}{dw} = \frac{MB}{GB}(r_1 - r_2) \qquad\qquad (6.33)$$

$$\frac{dy_G}{dw} = \frac{MB}{GB}(-\frac{n}{3}r_5 + \frac{n}{3}r_6) \qquad\qquad (6.34)$$

$$\frac{dT}{dw} = \frac{1}{GC_P}[(-\Delta H_1)(r_1 - r_2) + (-\Delta H_2)(r_4 - r_3) + \qquad\qquad (6.35)$$
$$(-\Delta H_3)r_5 + (-\Delta H_4)r_6]$$

From Equations 6.31 to 6.35, $r_1...r_6$ are defined as the rate constants, GB and MB are the mass flow rates of naphtha and recycle gas, Y_N, Y_P, Y_A and Y_G are the mole fractions of naphthenes, paraffins, aromatics and gases and n denotes the average number of carbon atom in molecules.

The rate constants and their dependence on temperature were obtained from (Smith, 1970). All the main reactions are reversible and their corresponding equilibrium constants were obtained from basic thermodynamic properties using the following:

$$\Delta F° = -RT \ln K \qquad\qquad (6.36)$$

For other reactions, however, it was necessary to calculate $\Delta F°$ as follows:

$$\frac{\Delta F^\circ}{T} = - \int_{T_0}^{T} \frac{\Delta H^\circ \, dT}{T^2} \qquad (6.37)$$

where ΔH° is the standard enthalpy change and ΔF° is the change in free energy associated with a particular reaction at a set temperature.

6.4.1.3 Numerical Integration

The system of first-order differential equations was solved by the method of Runge-Kutta. The Runge-Kutta numerical integration technique improves upon the accuracy of a simple integration by means of interpolation in which the non-linearity of the equation is compensated. PONA (paraffin, olefin, naphthene and aromatics) analysis gives the concentration of components on percent volume, these were converted to mole percentage and the method of calculation is given in Ansari (1981).

6.4.1.4 Results and Discussion

The test runs were carried out on a catalytic reforming unit and the results predicted by the model were compared with the plant data. Table 6.1 compares the overall results of reforming obtained by the model and the plant test runs. A number of model validations were needed through many sets of real process data to get these results. It may be observed from Table 6.1 that the model results differ with the plant test data for reformate and light gases. In this process model, this was not considered significant, as the main important variables were reactor temperatures and octane number. It may also be observed from the model results that low pressure reforming has an increasing effect on the production of aromatics. The operation at low pressure increases the fraction of feed energy, in terms of gross heat of combustion, contained in the desired product, reformate and hydrogen.

The process model was used to predict the product yields and operating conditions under different reactor inlet temperatures, feed changes and naphtha feed compositions. Two cases were studied at different operating conditions (feed rate changes) and the results are given in Tables 6.2 and 6.3.

Table 6.1. Plant test results compared to model yields

Variables	Model Results	Plant Test Results
Outlet Temperature, Reactor (R1)	450 °C	446 °C
Outlet Temperature, Reactor (R2)	464 °C	461 °C
Outlet Temperature, Reactor (R3)	485 °C	486 °C
Outlet Temperature, Reactor (R4)	492 °C	493 °C
Yield of Reformate C_5+ , Vol %	89.60	85.34
Production of Gases, Vol %		
Hydrogen	1.90	2.75
C_1	1.10	1.25
C_2	1.89	2.23

C_3	2.10	4.36
C_4	2.45	4.02
TOTAL	99.04	99.95
Composition of C_5+, Vol %		
Paraffins	45.30	47.52
Naphthenes	1.30	1.74
Aromatics	53.30	50.00
RON, Clear	97.80	97.50

Table 6.2. The prediction of product yields and reactor outlet temperatures for Case 1: Higher naphtha feed rate

Temperature and Yields	Model Results	Plant Test	Relative Error
Outlet Temperature, Reactor (R1)	452 °C	450 °C	0.4%
Outlet Temperature, Reactor (R2)	470 °C	468 °C	0.4%
Outlet Temperature, Reactor (R3)	498 °C	495 °C	0.6%
Outlet Temperature, Reactor (R4)	517 °C	510 °C	1.4%
Yield of Reformate C_5+ , Vol %	82.60	86.00	4.1%
Production of Gases, Vol %			
Hydrogen	4.00	3.80	5.3%
LPG	10.0	9.60	4.2%

Table 6.3. The prediction of product yields and reactor outlet temperatures for Case 2: Lower naphtha feed rate

Temperature and Yields	Model Results	Plant Test	Relative Error
Outlet Temperature, Reactor (R1)	448 °C	447 °C	0.2%
Outlet Temperature, Reactor (R2)	460 °C	458 °C	0.4%
Outlet Temperature, Reactor (R3)	487 °C	485 °C	0.4%
Outlet Temperature, Reactor (R4)	504 °C	500 °C	0.8%
Yield of Reformate C_5+ , Vol %	80.60	83.20	3.2%
Production of Gases, Vol %			
Hydrogen	2.70	2.85	5.5%
LPG	13.7	13.1	4.6%

These tables show that the maximum relative errors in the reactor outlet temperatures and product yields between the model prediction and plant were 1.4% and 5.5% respectively. At higher feed rates (higher liquid hourly space velocity), fewer catalyst sites will be available per unit of feed and so higher reactor temperatures will be required for the same conversion (Table 6.2). The net effect of an increase in feed rate at constant octane will be an improved platformate and hydrogen yield. Conversely low throughputs will allow time for hydrocracking reactions and so a minimum limit is nominally imposed on space velocity to avoid excessive yield loss. The net effect of decrease in feed rate is a reduction in both platformate and hydrogen yields (Table 6.3).

In Table 6.4, results of reactor temperatures derived from Smith's kinetic model were further compared with the previous work related to reactor outlet temperatures predictions on the fixed bed catalytic reforming processes. The temperature profiles prediction with Smith's model are more precise than the previous work with the maximum relative error within 0.9% as shown in Table 6.4.

Table 6.4. Comparison of reactor outlet temperatures (predicted by Smith's model) with other process models

Reactor Outlet Temperatures		R1 (°C)	R2 (°C)	R3 (°C)	R4 (°C)	Relative Error
Smith's Model	Prediction	450	464	485	495	0.9%
	Plant	446	461	486	493	
Lee *et al.* (1997)	Prediction	441	469	479	505	1.2%
	Plant	440	466	480	499	
Bommannan *et al.* (1989)	Prediction	434	475	489	-	4.9%
	Plant	434	453	476	-	

6.4.1.5 Conclusions

Smith's kinetic model has produced results, which are significantly closer to the plant test results as evident from Table 6.1. It was observed from the results of the model that was tested on the plant that low pressure reforming has an increasing effect on the production of aromatics and also on the quantity of hydrogen. In addition, the plant test runs reveal that operation at low pressure increases the fraction of feed energy, in terms of gross heat of combustion, contained in the desired products, reformate and hydrogen; however, because of the difference in selectivity and activity of different catalysts and unit configuration, it is not possible to predict precise simulation of a specific unit by this method.

6.4.2 Non-linear Control Algorithm for Reforming Reactors

6.4.2.1 Problem Formulation

The problem described in this section is applicable to square system where the manipulated and controlled variables are equal. For catalytic reforming reactors, if there are no constraints on any of the manipulated variables, the problem can be solved without using the optimization techniques with a unique set of reactor inlet temperatures.

Consider a process described by:

$$dy/dt = \mathbf{f}(\mathbf{y}, \mathbf{u}, \mathbf{d}, t) \tag{6.38}$$

$\mathbf{y}$ = output variable, $\mathbf{u}$ = input variable, d = measured disturbances, t = time

In GMC, the rate of change of the controlled variables is set equal to the PI term

$$dy / dt = \mathbf{k}_1 (\mathbf{y}_{sp} - \mathbf{y}) + \mathbf{k}_2 \int (\mathbf{y}_{sp} - \mathbf{y}) dt \qquad (6.39)$$

where $\mathbf{k}_1$ and $\mathbf{k}_2$ are diagonal matrices of the GMC parameters. Substituting Equation 6.38 into Equation 6.39 and rearranging yields the GMC control law:

$$\mathbf{f}(\mathbf{y},\mathbf{u},\mathbf{d},t) - \mathbf{k}_1 (\mathbf{y}_{sp} - \mathbf{y}) + \mathbf{k}_2 \int (\mathbf{y}_{sp} - \mathbf{y}) dt = 0 \qquad (6.40)$$

The velocity form of the control law of Equation 6.40 is given by:

$$\Delta \mathbf{f} (\mathbf{y},\mathbf{u},\mathbf{d},t) - \mathbf{k}_1 \Delta e - \mathbf{k}_2 e \Delta t = 0 \qquad (6.41)$$

where $e = \mathbf{y}_{sp} - \mathbf{y}$ and Δt is the sample time

Traditionally, more emphasis has been put into the development of steady-state process models as opposed to models accounting for process dynamics. Because GMC requires a dynamic model, approximate dynamics may be attached to a steady state model to replace dy/dt in Equation 6.38. Many process responses can be characterized by first-order dynamics (Cott, 1986).

Consider the following non-linear steady-state process model:

$$\mathbf{y}_{ss} = \mathbf{g}(\mathbf{u},\mathbf{d}) \qquad (6.42)$$

First-order dynamics may be approximated from an estimate of the open loop process time constant, *i.e.*,

$$f(\mathbf{y},\mathbf{u},\mathbf{d},t) = (\mathbf{y}_{ss} - \mathbf{y}) / \tau_p \qquad (6.43)$$

Substituting Equation 6.42 into the GMC control law, Equation 6.40, and rearranging gives:

$$\mathbf{y}_{ss} = \mathbf{y} + \mathbf{k}_1 \tau_p (\mathbf{y}_{sp} - \mathbf{y}) + \mathbf{k}_2 \tau_p \int (\mathbf{y}_{sp} - \mathbf{y}) dt \qquad (6.44)$$

Note that higher-order dynamics may be used in a similar manner. Equation 6.44 provides steady state targets for the controlled variables such that the pre-specified reference trajectories are followed. The next step is to calculate the manipulated variable values, $\mathbf{u}$, which will solve Equation 6.42 for $\mathbf{y}_{ss}$. If Equation 6.42 can not be solved explicitly for $\mathbf{u}$, then an iterative solution is required. The velocity form of GMC using a steady state model was obtained by making a discrete approximation and differencing Equation 6.44.

$$\Delta y_{ss,i} = \Delta y_i + \tau_p \, (\mathbf{k}_{1i} \, \Delta e_i + \mathbf{k}_{2i} \, e \, \Delta t) \tag{6.45}$$

Consider the first-order dynamics given by Equation 6.45, equating with Equation 6.38 and applying it to four catalytic Reforming reactor temperatures, we get the following set of equations representing the first-order dynamics with time constant.

$$\frac{dT_1}{dt} = (T_{1ss} - T_1) * \frac{1}{t_1} \tag{6.46}$$

$$\frac{dT_2}{dt} = (T_{2ss} - T_2) * \frac{1}{t_2} \tag{6.47}$$

$$\frac{dT_3}{dt} = (T_{3ss} - T_3) * \frac{1}{t_3} \tag{6.48}$$

$$\frac{dT_4}{dt} = (T_{4ss} - T_4) * \frac{1}{t_4} \tag{6.49}$$

For disturbance variable, F, which is a feed to the catalytic reforming reactors, the equation is written as follows:

$$\frac{dF}{dt} = (F_{ss} - F) * \frac{1}{t_F} \tag{6.50}$$

However, as the feed is constant and the composition of feed does not change very frequently, the term dF/dt in Equation 6.50 is considered to be zero.

The GMC control laws become:

$$T_{1ss} = T_1 + k_{1,1} \, t_1 \, (T_{1sp} - T_1) + k_{2,1} \, t_1 \int_0^t (T_{1sp} - T_1) dt \tag{6.51}$$

$$T_{2ss} = T_2 + k_{1,2} \, t_2 \, (T_{2sp} - T_2) + k_{2,2} \, t_2 \int_0^t (T_{2sp} - T_2) dt \tag{6.52}$$

$$T_{3ss} = T_3 + k_{1,3} \, t_3 \, (T_{3sp} - T_3) + k_{2,3} \, t_3 \int_0^t (T_{3sp} - T_3) dt \qquad (6.53)$$

$$T_{4ss} = T_4 + k_{1,4} \, t_4 \, (T_{4sp} - T_4) + k_{2,4} \, t_4 \int_0^t (T_{4sp} - T_4) dt \qquad (6.54)$$

where $T_{iss\,(i=1...4)}$ are the reactor target temperatures, $K_{1,j\,(j=1...4)}$ are the proportional gains of the temperature controller, $K_{2,j\,(j=1...4)}$ are integral gains of the temperature controller, $T_{isp\,(I=1...4)}$ are the reactor temperature setpoints and $T_{i\,(i=1..4)}$ are the current reactor temperatures.

Equations 6.51 to 6.54 provide steady-state targets for the controlled variable, weighted average inlet temperature (WAIT) such that the pre-specified reference trajectories are followed. The next step is to calculate the manipulated variable values, u, which will solve Equation 6.42 for y_{ss}.

It is important to note that for non-square systems or when Equation 6.40 has no solution, then the GMC control law must be constructed as an optimization problem which minimizes the difference between the process outputs and the reference trajectories defined in Equation 6.39. In the following section a constrained non-linear optimization problem is constructed for the catalytic reforming reactors as the system is non square and the two process constraints are present in the system such as heater tube skin temperature and fuel gas valve position. The general method described in Section 6.3 is followed and applied to the reactor section of catalytic reforming process.

6.4.2.2 Constrained Non-linear Optimization Problem

Choose: $T_1, T_2, T_3, T_4, \lambda^+_p, \lambda^-_p, \lambda^+_c, \lambda^-_c$

where $T_{i\,(i=1,...4)}$ are the heater inlet temperatures and λ^+_c, λ^-_c represent the variables departure from the chosen specification curves for the upper and lower constraints respectively.

To minimize:

$$J = W^-_p \lambda^-_p + W^+_p \lambda^+_p + W^-_c \lambda^-_c + W^+_c \lambda^+_c \qquad (6.55)$$

where W are weighting matrices with elements such that $w_{ii} \geq 0$, $w_{ij} = 0$, $i \neq j$. where i and j denote weighting factors on the ith and jth slack variables.

Subject to:

$$\frac{dT_1}{dt} + \lambda^+_{p_1} - \lambda^-_{p_1} = k_{11} (T_{1sp} - T_1) + k_{21} \int_0^t (T_{1sp} - T_1) \, dt \qquad (6.56)$$

$$\frac{dT_2}{dt} + \lambda^+_{P_2} - \lambda^-_{P_2} = k_{12} (T_{2sp} - T_2) + k_{22} \int_0^t (T_{2sp} - T_2)\, dt \tag{6.57}$$

$$\frac{dT_3}{dt} + \lambda^+_{P_3} - \lambda^-_{P_3} = k_{13} (T_{3sp} - T_3) + k_{23} \int_0^t (T_{3sp} - T_3)\, dt \tag{6.58}$$

$$\frac{dT_4}{dt} + \lambda^+_{P_4} - \lambda^-_{P_4} = k_{14} (T_{4sp} - T_4) + k_{24} \int_0^t (T_{4sp} - T_4)\, dt \tag{6.59}$$

In Equations 6.56 to 6.59, $T_{isp\ (i=1,2,3,4)}$ are the heater inlet temperature setpoints.

$$\frac{d\,(WAIT)}{dt} - \lambda^-_c \leq K_{1C} (WAIT_U - WAIT) \tag{6.60}$$

$$\frac{d\,(WAIT)}{dt} + \lambda^-_c \leq K_{2C} (WAIT - WAIT_L) \tag{6.61}$$

where $WAIT_U$ and $WAIT_L$ are the upper and lower bounds on the weighted average inlet temperature (WAIT) and K_{1C} and K_{2C} are pxp diagonal matrix selected using the GMC specification curves.

$$T_{iL} \leq T_i \leq T_{iU} \tag{6.62}$$

where T_{iL} and T_{iU} are the lower and upper constraints on input variables.

$$\Delta T_{iL} \leq T_i\,(t + \Delta t) - T_i\,(t) \leq \Delta T_{iU} \tag{6.63}$$

where ΔT_{iL} and ΔT_{iU} are the lower and upper constraints on the input movement variables.

$$\lambda^+_p \geq 0,\ \lambda^-_p \geq 0, \lambda^+_c \geq 0, \lambda^-_c \geq 0 \tag{6.64}$$

The inequality constraints on the controlled variables are the high and low limits placed on the system to represent plant-operating constraints.

The above optimization problem was applied to catalytic reforming reactors and the solution was obtained by using an equation-based non-linear optimization software RT-Opt. (real-time optimization) of AspenTech. The detailed discussions on RT-Opt. is given in Appendix D. The optimization problem was solved at every time step using the successive quadratic programming (SQP) algorithm in which the search direction is the solution of a quadratic programming problem. The optimizer allows the user to build and maintain a dynamic model and a steady-state optimization model within a single system. The optimizer also provides the on-line data validation and execution support for robust model-based solution. An alternative approach to solving non-linear optimization problem is to use the subroutine SOL/NPSOL described by Gill *et al.* (1986).

6.4.3 Non-linear Control Objectives and Strategies

6.4.3.1 Control Objectives

The main objective of non-linear multivariable control application to catalytic reforming reactor section is to control WAIT/octane at target values, and for a given octane and feed rate, maximize the yield of reformate consistent with the key operating constraints. The reformate octane is required to be maintained at a desired value established by the refinery's planning and scheduling section. Consistent with the overall refinery production requirements, the octane target is considered an optimum value based on gasoline blending requirements. As such there is a cost penalty associated with the octane target if it falls below or above the required specifications.

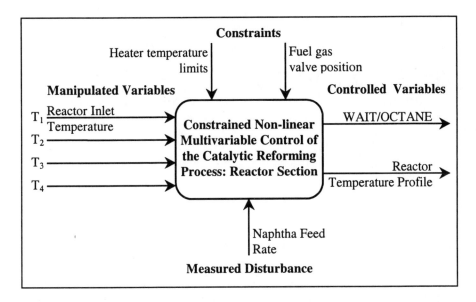

Figure 6.2. Non-linear multivariable control problem schematic for the catalytic reforming reactors

The control problem and its objective is further explained in Figure 6.2. Octane is primarily a function of weighted average inlet temperature, liquid hourly space velocity (LHSV), and feed properties. With LHSV determined by the feed rate to the unit, the manipulated variables for the octane/Wait control are the reactor inlet temperatures. The four controlled variables are the WAIT and the reactor temperature profile (the difference in the inlet temperature between the two reactors). Naphtha feed rate is the measured disturbance variable. The key constraint variables are heater tube skin temperatures and fuel gas valve positions.

6.4.3.2 WAIT/Octane Control Strategies

The WAIT/octane control strategy is shown in Figure 6.3. At the first level, a WAIT/Profile non-linear control function manipulates the four reactor inlet temperatures to maintain a target WAIT and reactor temperature profile. In this case, the profile is defined as the differences in inlet temperatures between reactors 2 versus 1, reactors 3 versus 2, and so on. As long as there are no constraints on any of the manipulated variables, the WAIT and profile targets can be satisfied with a unique set of inlet temperatures.

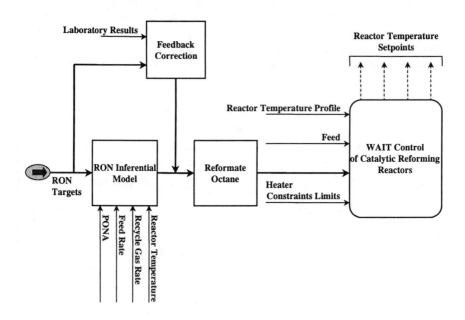

Figure 6.3. The WAIT/octane control strategies

When constraints are activated, the number of manipulated variables is reduced such that both the target WAIT and temperature profile can no longer be satisfied. In this case, the non-linear control function enforces constraint limits and maintains WAIT and temperature profile as close to the specified targets as possible. The non-linear multivariable control maintains the target WAIT as first priority and then relaxes the temperature profile as necessary to enforce constraint limits. The key constraint variables incorporated in the control strategy are heater tube skin temperatures and fuel gas valve positions.

A good control of WAIT and temperature profile is accomplished using a constrained non-linear model-based control methodology. This methodology provides a coordinated strategy for handling cases in which there are no active constraints on either the manipulated or controlled variables, as well as cases where constraints are activated and there is no longer a match between the number of manipulated and controlled variables. This approach also has the ability to dynamically compensate for the effects of disturbance variables, such as feed rate, on reactor inlet temperatures. Figure 6.4 shows a complete strategy of non-linear model-based multivariable control applied to the reactor section of the catalytic reforming process.

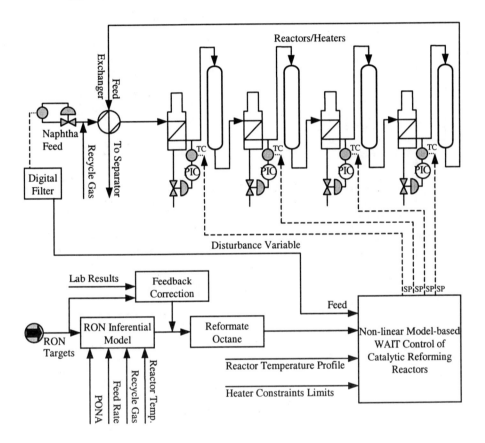

Figure 6.4. Non-linear model-based WAIT/octane control of the catalytic reforming reactors

6.5 Real-time Implementation

There are three important steps to implement the multivariable control applications in real-time. The first step is to follow the exact procedure and sequence of implementation. Figure 6.5 illustrates the implementation procedure of the non-linear model-based multivariable control on the reactor section of the catalytic reforming process. The second step is to configure the controller interface to distributed control system and finally the on-line controller tuning to meet the specified targets without process deviation and to gain the confidence of the operators.

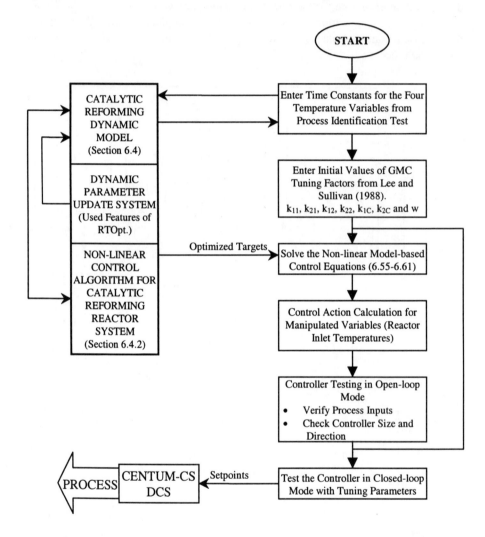

Figure 6.5. Flowchart illustrating the implementation procedure of non-linear model-based multivariable control on the catalytic reforming reactor section

6.5.1 Non-linear Controller Tuning

The time constants of the temperature controller (TC) residing in Distributed Control System (DCS) were estimated from open-loop test. The time to steady state was found to be about 90 minutes. The sampling frequency was taken once per minute. The desired trajectory of the process outputs were determined using the tuning rules by Lee and Sullivan (1988). The control application was first turned on and tested in an open-loop mode (*i.e.*, the outputs were suppressed). Open loop testing includes verifying process inputs and checking controller move sizes and direction. Next, the controls were tested in closed-loop mode using tuning constants gathered from the simulation model and step response test. Closed-loop testing was performed within a narrow operating range until confidence was gained. It was also noted that the form of optimization problem is well structured since a slack variable is added to each control law equation to ensure that a solution to the set of equations does in fact exists. It is important to note that if the control is implemented at a reasonable frequency, the solution of the NLP is very fast (three to four iterations) since the current control settings and slack variables provide a good initial estimate of the solution vector. Further, when the control system fails or turns itself off, the basic control system continues to function. In effect the operator does not have to do anything special to put the reactor on computer or to take it off.

As discussed in Section 6.3.2, two types of tuning parameters were used; the K_{1C} and K_{2C}, specifying the maximum speed of approach towards the constraint bounds and weighting factors w, reflecting the relative importance of the outputs and the constraints. The tuning parameters K_{1C} and K_{2C} can be compared with the move suppression factor, a term used in linear multivariable controller design such as dynamic matrix control (DMC). Changing the move suppression factor on a manipulated variable causes the controller to change the balance between movement of that manipulated variable, and error in the dependent variables.

The role of weighting factors **W** , is similar to the equal concern factor in linear multivariable control framework such as (DMC). In the WAIT/octane controller tuning analysis, the weighting factor for the WAIT (controlled variable) was assigned a value of 1 °C. It was observed that 1 °C on the WAIT was as important as 20 kpa on the fuel gas pressure. If the controller was unable to reach all the setpoints or limits, it would prioritize the errors in the ratio 1 °C = 20kpa.

6.5.2 Results and Discussions

An inferential model was used to predict the effects of key disturbance variables on reformate octane. The model was used in a feedforward mode to adjust the target WAIT based on changes in the disturbance variables. The model also provides an adjustment to the target WAIT for a change in target octane specified. The laboratory results were used to update the model based on the difference between the model prediction at the time the sample was taken and the actual measured value. Figure 6.6 shows the inferential model predicting reformate octane. The average deviation from the laboratory results is only $\pm$ 0.2 octane number, considering the repeatability of $\pm$ 0.3 octane number in laboratory results, this prediction can be considered very accurate. The details of the inferential modelling work was given in Chapter 4.

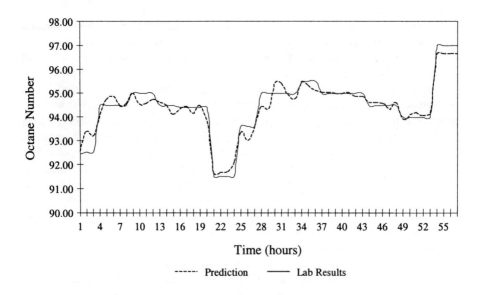

Figure 6.6. Inferential model predicting reformate octane

In Figure 6.7, the temperature profile of the reactors is shown. The details on how to control the reactor temperature profile were given in Section 6.4.4.2.

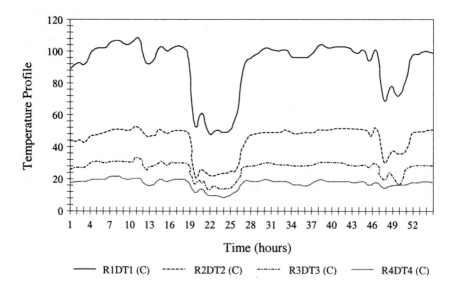

Figure 6.7. Catalytic reforming reactors temperature profile

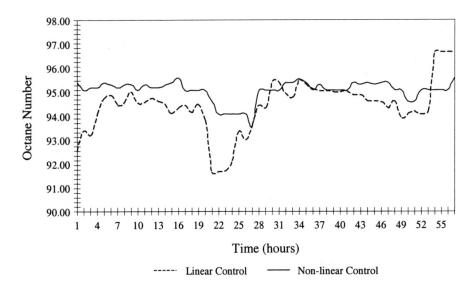

Figure 6.8. Comparison between linear and non-linear multivariable control for the reformate octane number

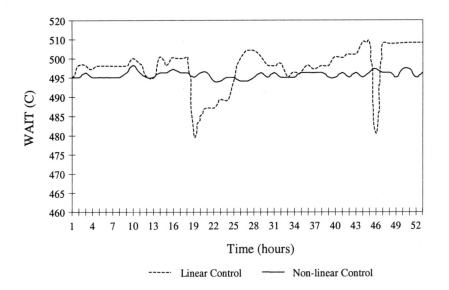

Figure 6.9. Comparison between linear and non-linear multivariable control for the catalytic reforming WAIT

Figures 6.8 and 6.9 show the comparison between linear model-predictive and non-linear multivariable control for octane number and WAIT. With linear control, the reformate octane deviates significantly from the target octane number

(95) when the feed rate is increased. The non-linear MVC reduced the octane deviation by 40 to 50% and gives a better response to disturbances in feed rate. Mark *et al.* (1987) has reported similar results on a catalytic reforming process using a linear model-predictive multivariable control system.

A detailed analysis of the two Figures 6.8 and 6.9 shows that when the feed rate increases and no other changes are made in the operation, the octane drops below target, the hydrogen to hydrocarbon ratio reduces to low operating range, the furnace duties increase above their rated capacity. Under linear control, the control system reduces WAIT to bring the furnaces back within constraint duties. In order to meet octane target at reduced WAIT, the only option is to increase pressure.

Figure 6.9 shows that the variation in WAIT is significantly low as compared to linear control system during increase in the feed rate. The non-linear MVC maintains a very good control of WAIT at 495 °C.

Figure 6.10 shows the reference trajectory for the constraints on the controlled variable (WAIT). The controller is free to follow GMC trajectory for the controlled variable as long as controlled variable is brought within limits or to a setpoint. The controller is obliged to keep the WAIT within the constraint and is allowed to follow any trajectory within these constraints. The controller has the maximum freedom to determine a trajectory that will require minimum manipulated variable movement and will be least sensitive to modelling errors. This concept is further elaborated by studying the two cases with constraints on manipulated variable and quality performance and constraint control.

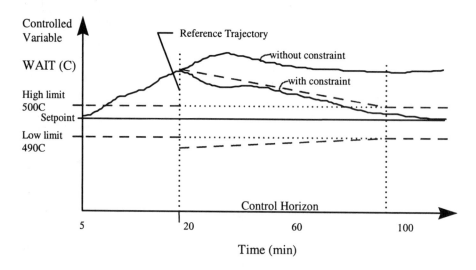

Figure 6.10. Reference trajectory of the process constraints for controlled variable

In the following discussions, two cases were studied to evaluate the GMC constraint formulation and to demonstrate the algorithm's ability to trade-off quality performance and constraint control.

Case 1: Constraints on Manipulated Variables
The catalytic reforming model was run for the unconstrained case using the
setpoints and initial GMC performance parameters shown in Table 6.5.

Table 6.5. Setpoint and GMC parameters

Output	Setpoints	ξ	τ (min)
WAIT	495 °C	3	90
ΔT_1	25 °C	3	90
ΔT_2	20 °C	3	90
ΔT_3	10 °C	3	90

For each output variable in the Table 6.5, the GMC controller parameters K_{1i} and
K_{2i} (the ith diagonal elements of K_{11} and K_{21}) from Equations 6.56 to 6.59 were
calculated by the following relationships:

$$K_{1i} = \frac{2\xi_i}{\tau_i} \tag{6.65}$$

$$K_{2i} = \frac{1}{\tau_i^2} \tag{6.66}$$

τ_i was estimated from the process identification test (plant test runs) and ξ_i was
selected from the GMC performance specification curve (Lee and Sullivan,1988).
This pair of parameters have explicit effects on the closed-loop response of output
variable. The parameter ξ_i determines the shape of the closed-loop response while τ_i
determines the speed of the response (large τ_i means slow response). The GMC
design parameters were selected as discussed in Section 6.3.2.

For the unconstrained case, the controlled variable does not reach its setpoint value.
In order to examine the performance of the system with the GMC constraint
algorithm, constraints were added to rate of change of four input (manipulated)
variables as shown in Table 6.6.2.

Table 6.6. Variables and process constraints

Variables	Constraints Lower Bound	Constraints Upper Bound	Units
Temp (T_1)	460	460	°C
Temp (T_2)	480	480	°C
Temp (T_3)	495	495	°C
Temp (T_4)	500	500	°C
WAIT	480	495	°C
$\Delta T_{i(I=1,2,3)}$	0.1	0.5	°C

The rate of change constraint on the reactor temperature was significant enough to reduce the response time of the reactor temperature. The controlled variable (WAIT) comes to its setpoint value after an undershoot as shown in Figure 6.11.

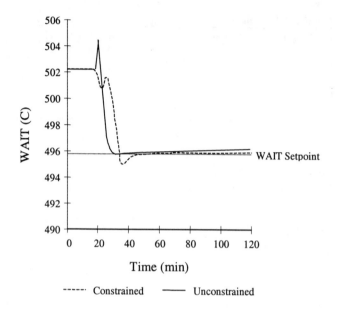

Figure 6.11. Process constraints on manipulated variables

Case 2: Quality Performance and Constraint Control
In Case 2 the simulation model was run to demonstrate the algorithm's ability to trade off quality performance and constraint control. The active constraint for WAIT is given in Table 6.6. Figure 6.12 shows that when the slack variable weightings in the objective function favour constraint handling (WAIT constraint prioritized), the constraints are satisfied at the expense of quality performance. Similarly if quality is important, then the constraints are relaxed to achieve the required WAIT at setpoint value. This case can also be understood through the inspection of Equations 6.60 and 6.61 and the slack variables weighting factors in the objective function given by Equation 6.55.

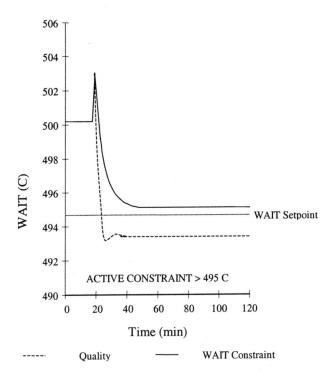

Figure 6.12. Trade-off between setpoint achievement and constraint satisfaction

6.6 Conclusions

A constrained non-linear optimization strategy for handling the constraints has been developed and applied in real-time to the catalytic reforming reactor section. An inferential model predicting the octane number was also developed and integrated with multivariable controller forming a closed loop WAIT/octane control. A dynamic model of the catalytic reforming process was developed and used to provide target values for the reactor inlet temperatures. It has been shown that non-linear multivariable control provided better disturbance rejection compared to traditional linear model-based control (Figures 6.8 and 6.9). The non-linear MVC reduced the octane deviation by 40%. Also it has reduced the variation in WAIT and maintained a very good control of WAIT at 495°C. The same model is used for optimization and control, minimizing the modelling errors due to process/model mismatch. The main contribution of this work is to combine non-linear multivariable controller with non-linear constrained optimization models and its application to a highly non-linear catalytic reforming process

In the optimization strategy, maximum rates of approach towards the constraint bounds are specified. Slack variables are introduced both for the output performance and for the constraints. These slack variables are measurements of the performance degradation and the potential constraint violation. The weighting factors of the objective function reflect the relative importance of the process outputs the constraints and control manipulations. While the application of a NLP algorithm given in Section 6.3.1 at every sampling time is essential, this strategy is applicable to non-square systems and those systems where GMC control has no solution.

Through the optimization formulation, it has been shown that the merit of this technique is the degree of flexibility in establishing the proper balance between the violation of the constraint variables and the deterioration of the control performance coupled with the ability to pre-define the response trajectories for both the controls and constraints.

The process model played a vital role in this strategy. It was used to make the outputs follow their trajectories in Equations 6.56 to 6.59 and to avoid the constraint violations in Equations 6.60 and 6.61. It shows that the accuracy of the model is very important for the success of this strategy. As there is always a process/model mismatch, implementing a reliable, on-line estimator for parameters and disturbances is essential for the success of this strategy. An alternative approach for model inaccuracy problem is to integrate constraint-handling strategies with the process/model mismatch compensation algorithm. This integration will be demonstrated in the following chapter.

CHAPTER 7

NON-LINEAR MULTIVARIABLE CONTROL OF A FLUID CATALYTIC CRACKING PROCESS

In this chapter, a non-linear constrained optimization strategy is proposed and applied to a fluid catalytic cracking reactor-regenerator section. A dynamic parameter update algorithm was developed and used to reduce the effect of larger modelling errors by regularly updating the selected model parameters. This algorithm is capable of adapting the model parameters in a non-linear model. This chapter also presents and discusses the dynamic process models identified from open-loop step test data. The constrained non-linear optimization algorithm and strategies were tested in real-time on a fluid catalytic cracking reactor-regenerator section and the results were compared to linear multivariable controller such as dynamic matrix control (DMC).

A non-linear dynamic model of a fluid catalytic cracking process was used for dynamic analysis of the plant and non-linear multivariable control system. The model realistically simulates the riser-reactor, one stage regenerator by assembling the mass and energy balances on the system of reactions and are then used for non-linear multivariable control application on the reactor-regenerator section. The model results were tested in real-time application and the results were used to improve the dynamic response of the model and to provide the initial values for the non-linear control system design. The constrained non-linear multivariable controller controls the flue gas oxygen concentration to a given setpoint and the regenerator bed temperature is controlled to within high and low zone limits. All other controlled variables are controlled to one-sided zone limits. These limits set bounds on the unit's throughput and cracking severity.

7.1 Introduction

The Fluid Catalytic Cracking (FCC) unit converts gas oils into a range of hydrocarbon products of which gasoline is the most valuable. In any refinery, the amount of low market-value feedstocks available for catalytic cracking is considerable, and a typical FCC unit's ability to produce gasoline from low market value feedstocks gives FCC unit a major role in the overall economic performance of a refinery; therefore it is a prime candidate for any kind of advanced control applications. There is, however, even more challenging task ahead of a process control engineer. With its internal feedback loop created by the circulating catalyst and its complex dynamic responses, the FCC unit is a truly fascinating process to observe and to analyze. The challenge stems from the characteristics and from the fact that controlled variables must be kept not so much at target but against constraints, where the process often behaves highly non-linearly. Economic objectives on manipulated variables, such as the need to maximize either charge rate or cracking severity, or both, together with the operating constraints further complicate the control problem (Prett and Gillette, 1979).

For the reactor-regenerator system, several dynamic models have been presented in the literature (Kurihara (1967); Zheng (1994); Moro and Odloak (1995)). Although the engineering principles of all FCC models are the same, the dynamic effects vary depending on the geometric configuration of the system. The scope of dynamic modelling in this chapter is not to describe the complex kinetics of the cracking reactions or the intricate hydrodynamics of the fluidized regenerator beds. The scope is to include in the model enough details to capture the control relevant dynamics, without sacrificing important aspects such as the description of non-linearities and interactions. The FCC unit is a typical example of a constrained multivariable process, where various predictive controllers have been commercially applied (Caldwell and Dearwater, 1991) with reported good results. In the majority of the practical cases, the most profitable operating point lies on the interception of several FCC constraints. The predictive controllers are capable of including constraints in the formulation of the control law. In the Dynamic Matrix Control (DMC) approach represented by (Cutler and Ramaker, 1980) and model algorithm control (Rouhani and Mehra, 1982) constraints can be added to the predicted error equations. The controller calculates the manipulated variables that minimize the errors in a least squares sense. Only quality constraints can be handled by this strategy. In the quadratic dynamic matrix control (QDMC) approach, the control problem is formulated as a quadratic-programming problem. The cost function is the square of the distance from the predicted to the reference trajectories. Constraints on the controlled and manipulated variables can be explicitly included and rigorously attended. Therefore, both linear programming (Chang and Seborg, 1983) and quadratic programming (Little and Edgar, 1986) have been applied in model-predictive control for processes with linear constraints. Internal Model Control (Garcia and Morari, 1982, 1985), as a control framework, is also a linear model-based control. Linear programming (Brosilow et al., 1984) and quadratic programming (Ricker, 1985) were also applied for constraint treatment in a similar manner to that applied for the model-predictive control algorithms.

Economou et al. (1986) extended IMC to non-linear lumped parameter systems by an operator approach. The key issue was inversion of the non-linear process model in the control law, and a Newton-type method was adopted. Li and Biegler (1988) have recently extended this operator approach to deal with linear input and state variable constraints by a successive quadratic programming (SQP) strategy.

In a different approach, a control framework for both linear and non-linear systems, generic model control (GMC) has been developed in the time domain (Lee and Sullivan, 1988). The control law employs a non-linear process model directly within the controller (Lee, 1991). Also, an integral feedback term is included, such that the closed-loop response exhibits zero offset. GMC was further extended to compensate for process/model mismatch (Lee *et al.*, 1989) and deadtime compensation (Lee *et al.*, 1990), but the strategies for constraint handling within the GMC framework have not yet been explored extensively, and have not been applied to complex processes in petroleum refineries. Brown *et al.* (1990) have applied these control strategies to two non-linear simulated systems. The first system was a forced circulation single-stage evaporator and the second system was a stirred tank recator from the paper of Li and Biegler (1988). The authors did not consider process-model mismatch into the control strategy. Their work did not address the non-linear controller implementation issues such as real-time application and on-line tuning of the controllers. The authors also did not address the complexities of handling the multiple constraints and process interaction, which are the characteristics of the FCC process addressed in this chapter. Also, the authors did not compare the simulation results with the existing linear model-based control. The latest development in neural network model-predictive control by Ishida and Zhan (1995) and the work of Spangler (1994) and Keeler *et al.* (1996) was discussed in detail in Chapter 6.

This chapter provides details of the application of the constraint non-linear multivariable control and solves a non-linear optimization problem on the reactor-regenerator section of a fluid catalytic cracking unit. The main contribution of this work is to incorporate directly the non-linear FCC process dynamic model with dynamic parameter update system within the non-linear multivariable control algorithm and to develop a non-linear constrained optimization strategy and its application to a highly non-linear FCC process with multiple constraints and process interaction. In addition dynamic process models were identified from open-loop step test data and a linear model-predictive control algorithm was developed in MATLAB® (The Mathworks, Inc., 1993) using the features from established model-based control algorithms such as dynamic matrix control and model algorithm control. In simulation studies, the linear controller results were compared to non-linear controller and it has been shown that non-linear controller outperforms the linear controller with regards to its ability to control the process with inverse response. In real-time application, the constrained non-linear optimization problem was constructed using simplified FCC process model with a dynamic parameter update system. The non-linear controller was tested and compared to a DMC type controller built with the dynamic models obtained from process identification tests. The main advantage of the proposed non-linear development is that a single-time step control law resulted in a much smaller-dimensional non-linear programme as compared to other previous methods. The other advantage of this approach is to minimize the time spent to do the plant testing as it uses non-linear process models and therefore needs much less maintenance and re-identification than the traditional linear controller that often have to be re-identified as the process moves into a new region of operation.

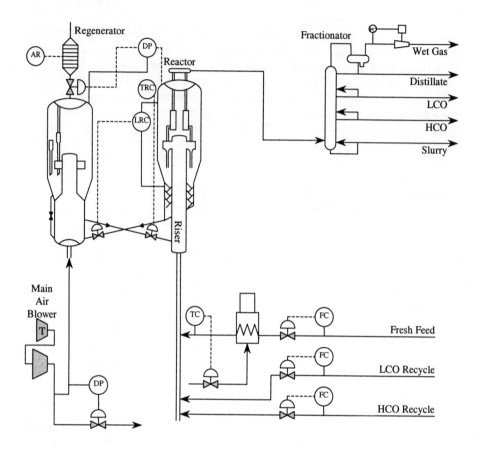

Figure 7.1. An overview of the fluid catalytic cracking process

7. 2 FCC Process Control Overview

7.2.1 Process Description

An overview of the FCC process is shown in Figure 7.1. The FCC process converts heavy oils into lighter and more valuable products. The unit consists of a reactor-regenerator section, main fractionator, and gas processing facilities. Fresh feed is heated in a fired heater and combined with regenerated catalyst in the reactor riser. Reaction temperature is controlled at 530 °C by manipulating the flow of hot regenerated catalyst to the reactor-riser. The catalyst and reacted hydrocarbon vapours flow up the riser and are separated in the reactor cyclone. Catalyst circulation is achieved by burning off the coke deposit in a fluidized bed inside the regenerator. A steam turbine driven air blower supplies the oxygen needed to burn the coke deposit. The spent catalyst is held up in a small fluidized bed in the stripping section of the reactor before being returned to the regenerator.

Combustion air flow is controlled by adjusting the blower blade angle. The flue gas from the regenerator is sent to a waste heat boiler before being discharged to the atmosphere. The FCC unit is operated in full burn mode and the flue gas, therefore, contains a small amount of excess oxygen, typically 1%. Vapour products from the reactor are sent to the bottom of the main fractionator where various boiling point fractions are withdrawn such as distillate, light cycle oil (LCO) and heavy cycle oil (HCO) *etc*. Feed rate, reaction temperature, and regenerator air rate are the main operating variables used for control. There are typically two to four feeds which include fresh feeds from the crude and vacuum units, and recycle feed from the main fractionator. All of the feeds may be used for control, however, the process may respond the same way for two or more feeds.

7.2.2 Economic Objectives and Non-linear Control Strategies

The economic objective of FCC unit is often to maximize feed rate at constant riser outlet temperature. A more sophisticated approach has been applied here by using the dynamic model of the process to calculate economic optimal values of the key operating variables. Optimal values are calculated for the feed rates, riser temperature, and feed preheat temperature and are enforced subject to operating constraints on the process. In addition, dynamic model parameter update algorithm is also used to update some key operating variables.

The non-linear multivariable control strategies have been applied to FCC because of its ability to handle the multivariable interactions and constraints, its ease of configuration and implementation, adaptability and its robustness. Although it can be applied to complete FCC process, this chapter will focus only on the most complex multivariable controls, those for the reactor-regenerator section. In general, the FCC reactor-regenerator non-linear control strategies is divided into two modules:

- Regenerator Loading Control Module
- FCC Severity Control Module.

The regenerator loading control module maintains the oxygen in the flue gas by manipulating the air flow rate to the regenerator. The primary control variable is the flue gas composition subject to temperature constraints. In complete combustion operation, the loading control regulates flue gas %O_2. In partial combustion operation, flue gas CO/CO_2 ratio is controlled. On-line analysis is used for flue gas composition control. The control compensates for changes in feed rate, recycle flow rate, riser outlet temperature and feed temperature. Figure 7.2 shows the non-linear multivariable control system of controlled and manipulated variables for the FCC unit with various constraints on the system.

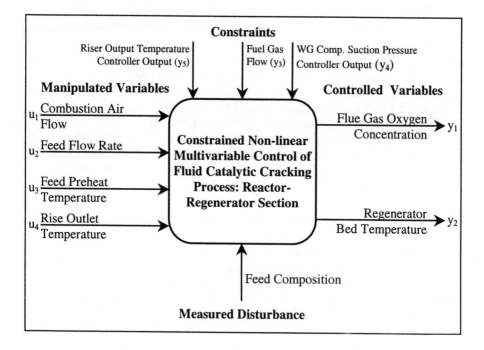

Figure 7.2. Non-linear multivariable control problem schematic for the catalytic cracking reactor-regenerator system

The FCC severity control module operates the reactor-regenerator at the most desirable point while observing the various process constraints. Optimum operation of the FCC units usually occurs at multiple constraints (Martin *et al.*, 1985). The module manipulates the riser outlet temperature, feed flow rate and feed temperature to control the unit within its constraints while trying to satisfy a specified operating objective. Figure 7.3 shows the non-linear multivariable control strategies on the FCC reactor-regenerator system. In this control strategy, the severity and loading control have been combined and solved simultaneously. The combined strategy also works well if the regenerator becomes air limited.

7.3 Dynamic Model of the FCC Process

7.3.1 Model Development

The FCC unit consists of a cracking reactor where the desired reactions include cracking of high boiling gas oil fractions into lighter hydrocarbons and the undesired include carbon formation reactions, and a regenerator, where the carbon removal reactions take place. Detailed discussions on the features of the FCC unit are given in Huq *et al.* (1995) and McFarlane *et. al.* (1993).

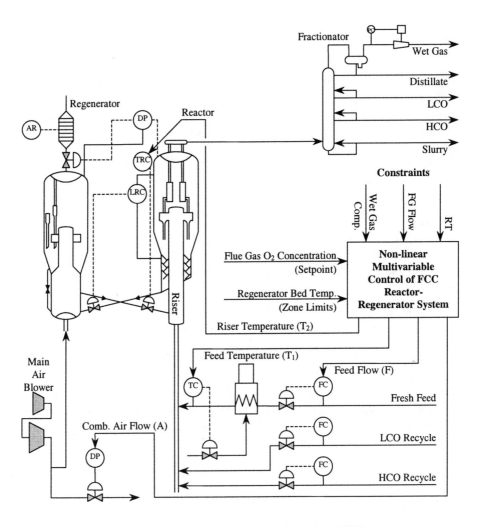

Figure 7.3. Non-linear multivariable control strategies on the FCC reactor-regenerator system

Under the following standard modelling assumptions:

1. Well-mixed reactive catalyst in the reactor.
2. Small-size catalyst particles.
3. Constant solid hold-up in reactor and regenerator.
4. Uniform and constant pressure in reactor and regenerator.
5. Full combustion was considered instead of partial combustion.

The non-linear process model considered here takes the following simplified form of the model given by Denn (1986). The scope of this simplified process model in this work is to include in the model enough details to capture the control relevant dynamics, without sacrificing the non-linearities and interactions of the FCC process.

This dynamic model is a short-cut approach to modelling and simulating the FCC process. Martin *et al.* (1985) used a rigorous model to simulate the FCC process. Zheng (1994) and Moro and Odloak (1995) used extensive modelling techniques to simulate the FCC reactor-regenerator system. These authors translated the economic objectives of the FCC system into a linear combination of the controlled and manipulated variables. However, these models were not used for non-linear multivariable control applications. The application presented in this chapter is unique in a sense that it used the shorter version of the FCC dynamic model for simulation purposes and then the same model was used for non-linear multivariable control applications.

7.3.1.1 Riser and Reactor Section

The coke generated by the cracking reaction is usually known as catalytic coke. The balance of catalytic coke in the reaction section (riser-reactor) can be written as:

$$V_{ra} \frac{dC_{cat}}{dt} = -60 \, F_{rc} \, C_{cat} + 50 \, R_{cf} \tag{7.1}$$

$$V_{ra} \frac{dC_{sc}}{dt} = -60 \, F_{rc} \, (C_{rc} - C_{sc}) + 50 \quad R_{cf} \tag{7.2}$$

where C_{cat}, C_{sc}, C_{rc} denote the concentrations of catalytic carbon on spent catalyst, the total carbon on spent catalyst, and carbon on regenerated catalyst. V_{ra} denotes the holdup of the reactor, F_{rc} denotes the flow rate of catalyst from the reactor to the regenerator and R_{cf} denotes the reaction rate.

The dynamics of the cracking reaction in the riser is negligible when compared to dominant time constants of the system. This is justified by the small residence time (about 2 seconds) of the hydrocarbons and the catalyst in the riser. This leads to the following energy balance equation in the riser:

$$V_{ra} \frac{dT_{ra}}{dt} = 60 \, F_{rc} \, (T_{rg} - T_{ra}) + 0.875 \, \frac{S_f}{S_c} \, D_{tf} \, R_{tf} \, (T_{fp} - T_{ra})$$

$$+ 0.875 \, \frac{-\Delta H_{fv}}{S_c} \, D_{tf} \, R_{tf} \tag{7.3}$$

$$+ 0.50 \, \frac{-\Delta H_{cr}}{S_c} \, R_{oc}$$

where T_{ra} and T_{rg} denote the temperatures in the reactor and the regenerator, D_{tf} is the density of total feed, S_f and S_c denote specific heats, R_{oc} denote reaction rate, T_{fp} denote the inlet temperature of the feed in the reactor and R_{tf} denote the total feed rate. ΔH_{fv} and ΔH_{cr} are the heat of feed vaporization and the heat of reaction.

7.3.1.2 Regenerator Section

The main equations of the model for the regenerator are:

Coke balance in the regenerator

$$V_{rg} \frac{dC_{rc}}{dt} = 60 \, F_{rc} \, (C_{sc} - C_{rc}) - 50 \, R_{cb} \tag{7.4}$$

Energy balance in the regenerator

$$V_{rg} \frac{dT_{rg}}{dt} = 60 \, F_{rc} \, (T_{ra} - T_{rg}) + 0.5 \, \frac{S_a}{S_c} \, R_{ai} \, (T_{ai} - T_{rg}) \tag{7.5}$$

$$+ 0.50 \, (\frac{-\Delta H_{rg}}{S_c}) \, R_{cb}$$

The analytic expressions for the reaction rates R_{cf}, R_{oc} and R_{cb} were obtained from (Denn, 1986):

$$R_{cf} = \frac{k_{cc} \, V_{ra} \, P_{ra}}{C_{cat} \, C_{rc}^{0.06}} \exp \left\{ \frac{-E_{cc}}{R(T_{ra} + 460.0)} \right\} \tag{7.6}$$

$$R_{oc} = \frac{k_{cr} \, V_{ra} \, P_{ra} \, R_{tf} \, D_{tf} \, K_{cr}}{R_{tf} + V_{ra} \, P_{ra} \, K_{cr}} \tag{7.7}$$

$$K_{cr} = \frac{k_{cr}}{C_{cat} \, C_{rc}^{0.15}} \exp \left\{ \frac{-E_{cr}}{R(T_{ra} + 460.0)} \right\} \tag{7.8}$$

$$R_{cb} = \frac{R_{ai} \, (21 - O_{fg})}{200} \tag{7.9}$$

$$O_{fg} = 21 \exp \left\{ \frac{-\dfrac{V_{rg} \, P_{rg}}{R_{ai}}}{\dfrac{10^6}{4.76 R_{ai}^2} + \dfrac{100}{k_{or}} \exp \left\{ \dfrac{-E_{or}}{R(T_{rg} + 460.0)} \right\} C_{rc}} \right\} \tag{7.10}$$

where k_{cc}, k_{cr} and k_{or} are pre-exponential kinetic rate constants, E_{cc}, E_{cr}, E_{or} are activation energies, O_{fg} is the oxygen in flue gas, R_{tf} is the total feed flow and R_{ai} is the air rate.

The values of the process parameters and the corresponding steady-state values are given in the Appendix E. In the model considered here the coke formation rate given by Equation 7.6 and the cracking reaction rate given by Equation 7.7 are directly affected by the gas oil composition. The two important parameters involved in Equation 7.6 were updated using the dynamic parameter update system described in Section 7.5.

The system of first-order non-linear ordinary differential equations was solved by the method of Runge-Kutta. The Runge-Kutta numerical integration technique improves upon the accuracy of a simple integration by means of interpolation in which the non-linearity of the equation is compensated. The method used to solve these equations is the same as discussed in Section 6.4.

7.3.2   Results and Discussion

In view of control, a process model should represent the dynamic behaviour of the system inside its operating range. This means that the dominant time constants and the gains of both plant and model must match closely. In this work a standard set of parameters is used, and biases are added to the model results to allow comparison with the plant data. Large excursions on the plant temperatures are allowed in the model with a slightly larger time constant for the regenerator but with very close gains in all cases. In the comparison test a step change of about 5% was introduced on the setpoint of the air flow controller, and the air flow pattern is shown in Figure 7.4.

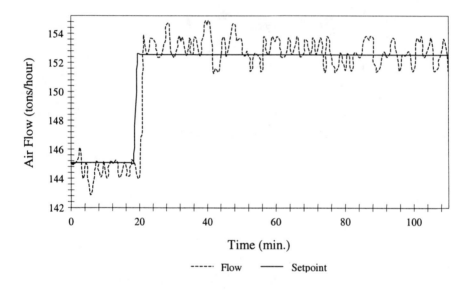

Figure 7.4. Air flow pattern on the 5% change in the setpoint of the air flow controller

The step change in the simulation model is of the same magnitude and value as the real disturbance. Figures 7.5 and 7.6 show the plant and model responses for the regenerator and the riser temperatures with the changes in the air flow rate.

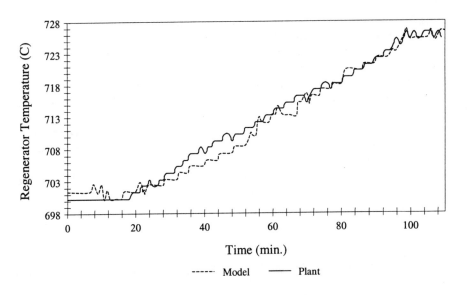

Figure 7.5. The plant and model responses for the regenerator temperature with the changes in air flow rate

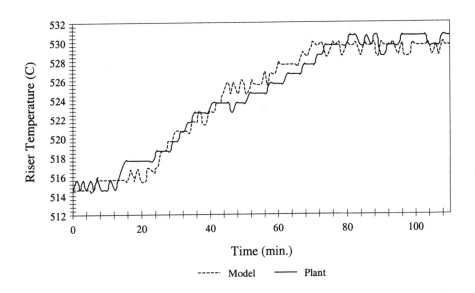

Figure 7.6. The plant and model responses for the reactor riser temperature with the changes in the air flow rate

The model responses are quite satisfactory. This shows that, for the control purposes the model can represent the FCC dynamics with an adequate degree of accuracy. In real-time application, for some of the controlled variables, the model step responses were used in the non-linear multivariable controller design, without extensive process identification test. However, the process identification test was performed for the development of linear controller design such as dynamic matrix control (DMC) and discussed in details in Section 7.6.2.

Three cases were studied at various feed rates and feed quality. Case 1 uses fresh feed without LCO and HCO recycles, Case 2 uses fresh feed with LCO recycle and Case 3 is the combination of all three feeds. Based on these feeds, the test runs were carried out on the FCC unit and the results predicted by the model were compared with the plant data. Table 7.1 compares the overall results of FCC process obtained by the model and the plant test runs. A number of model validations were needed through many sets of real process data to get these results. The Table 7.1 shows that the maximum relative errors in the riser outlet and regenerator bed temperatures between the model prediction and the plant were 0.3% and 0.5% respectively, whereas the average relative error in flue gas oxygen is 6.8%.

Table 7.1. Comparison of model prediction with the plant results

Variables		Units	Case 1	Case 2	Case 3	Relative Error %
Riser Outlet	Prediction	°C	515	520	530	0.3
Temperature	Plant	°C	517	521	529	
Regen Bed	Prediction	°C	705	710	720	0.5
Temperature	Plant	°C	708	712	725	
Air Flow Rate	Prediction	tons/hr	145	148	150	2.0
	Plant	tons/hr	147	150	155	
Feed Flow Rate	Prediction	m³/hr	100	105	110	1.6
	Prediction	m³/hr	102	107	111	
Oxygen in Flue	Prediction	%vol.	1.2	1.5	1.8	6.8
Gas	Plant	%vol.	1.3	1.6	1.9	

7.4 Non-linear Control Algorithm for FCC Reactor-Regenerator System

Select: $A, F, T_1, T_2, \lambda^+_p, \lambda^-_p, \lambda^+_c, \lambda^-_c$

where A is a combustion air flow, F is a feed flow, T_1, T_2 are the feed pre-heat and riser outlet temperatures respectively.

To Minimize:

$$J = \mathbf{W}^-_p \lambda^-_p + \mathbf{W}^+_p \lambda^+_p + \mathbf{W}^-_c \lambda^-_c + \mathbf{W}^+_c \lambda^+_c + \Delta \mathbf{u}_{i\,(i=1,2,3,4)} \qquad (7.11)$$

where $\mathbf{W}$ are weighting matrices with elements such that $w_{ii} \geq 0$, $w_{ij} = 0$, $i \neq j$ and i and j denote weighting factors on the ith and jth slack variables. J is an objective function to be minimized. $\mathbf{u}_i$ are the manipulated variables as defined above. In addition, the positive values of λ^-_p or λ^+_p represent the difference between output response and the GMC reference trajectory. This difference is a measurement of the control performance degradation. λ^-_p or λ^+_p, are defined to express the systems negative offset and positive offset from the pre-specified response trajectories.

Positive values of λ^-_c or λ^+_c, are the difference between the actual rates of changes of the constraints beyond the specified maximum rate of approach towards the constraint bounds. The larger they are, the more likely that the constraints will exceed their bounds in the future. Thus they are the measurement of the potential constraint violation. The values of these slack variables will be chosen in the proposed optimization problem that will allow the control algorithm to "trade-off" performance versus constraint violation.

The slack variable weighting factors, W, reflect the relative importance of the outputs and constraints. In selecting the value of W, one should consider the effect of the different scales of the outputs and constraints. The selection of W is also useful for differentiating between soft constraints and hard constraints. For hard constraints, where the constraint control is more important than the quality of control, W should be selected as:

$$W^+_c, W^-_c >>> W^+_p, W^-_p \tag{7.12a}$$

while for soft constraints the relative values of W_c to W_p reflect the trade-off. If the quality of control is more important than the constraint control, then W can be selected as:

$$W^+_p, W^-_p >>> W^+_c, W^-_c \tag{7.12b}$$

for which the functional constraints virtually have no effect on the closed-loop responses of the system.

The ability to assign independent weightings on selected constraints and quality variables allows the controller to prioritize the process response. If the control of the ith output is more important than the control of the jth output, then

$$W^+_{pii} > W^+_{pjj} \; ; \; W^-_{pii} > W^-_{pjj} \tag{7.13}$$

Subject to:

$$\frac{dy_1}{dt} + \lambda^+_{p_1} - \lambda^-_{p_1} = k_{11}(y_{1sp} - y) + k_{21} \int_0^t (y_{1sp} - y)\, dt \tag{7.14}$$

Equation for flue gas oxygen concentration

$$\frac{dy_2}{dt} + \lambda^+_{p_2} - \lambda^-_{p_2} = k_{12}(y_{2sp} - y) + k_{22} \int_0^t (y_{2sp} - y)\, dt \tag{7.15}$$

Equation for regenerator bed temperature

$$\frac{d\,y_2}{dt} - \lambda\,\dot{}_c \le K_{1C}\,(y_{2U} - y) \tag{7.16}$$

$$\frac{d\,y_2}{dt} + \lambda\,\dot{}_c \le K_{2C}\,(y_2 - y_{2L}) \tag{7.17}$$

Equations for regenerator bed temperature constraints

$$\frac{d\,y_3}{dt} + \lambda\,^+_c \le K_{3C}\,(y_{3U} - y_3) \tag{7.18}$$

Equation for fuel gas flow

$$\frac{d\,y_4}{dt} + \lambda\,^+_c \le K_{4C}\,(y_{4U} - y_4) \tag{7.19}$$

Equation for wet gas compressor suction pressure output

$$\frac{d\,y_5}{dt} + \lambda\,^+_c \le K_{5C}\,(y_{5U} - y_5) \tag{7.20}$$

Equation for riser outlet temperature controller output

$$A_L \le A \le A_U \quad \text{(input constraint)} \tag{7.21}$$

$$\Delta A_L \le \Delta A\,(t +\Delta t) - \Delta A\,(t) \le \Delta A_U \quad \text{(input movement constraint)} \tag{7.22}$$

$$\lambda^+_p \ge 0,\ \lambda^-_p \ge 0,\ \lambda^+_c \ge 0,\ \lambda^-_c \ge 0 \tag{7.23}$$

The inequality constraints on the controlled variables are the high and low limits placed on the system to represent plant operating constraints. In all FCC units operating in full burn mode, coke combustion is essentially complete and increasing the flow of combustion air to the regenerator can only raise the flue gas oxygen concentration. All other controlled variables are controlled to one-sided zone limits. These limits set bounds on the units throughput and cracking severity.

The optimization problem can be solved using a non-linear constrained optimization algorithm. The form of the optimization problem is well structured since a slack variable is added to each control law equation to ensure that a solution to the set of equations does exist. If the control is implemented at a reasonable frequency, the solution of the NLP is very fast since the current control settings and slack variables provide a good initial estimate of the solution vector.

The above optimization problem was applied to FCC reactor-regenerator section and the solution was obtained by using the Optimization toolbox of MATLAB®. The toolbox contains many commands for the optimization of general linear and non-linear functions. The optimization toolbox contains **CONSTR.M** file that finds the constrained minimum of a function of several variables. The program uses sequential quadratic program (SQP) method in which the search direction is the solution of a quadratic-programming problem. From the Non-linear Control Design (NCD) toolbox, a non-linear optimization file **NLINOPT.M** was called to run the optimization algorithm, which in return calls a routine which converts lower and upper bounds into constraints used by the optimization. This technique to solve the non-linear control algorithm is easy to use and implement as compared to equation based non-linear optimization software RT-Opt. (real-time optimization) of Aspen Tech described and used in Chapter 6. In RT-Opt., more time and efforts are required to modify the program template in the software to make it specific to the required application.

An alternative approach to solving non-linear optimization problem is to use the subroutine SOL/NPSOL described by Gill *et. al.* (1986). NPSOL is a selection of FORTRAN subroutines designed to solve the non-linear programming problem: the minimization of a smooth non-linear function subject to a set of constraints on the variables.

7.5 Dynamic Model Parameter Update

7.5.1 Introduction

In a model-based controller, including generic model control (GMC), there is an element of mismatch between the model and the true process. This process-model mismatch leads to deterioration in control performance. There are two types of model mismatch: structured mismatch occurs when the process and the model are of a different nature (e.g., first-order/second-order, or linear/non-linear); parametric mismatch occurs when the numerical values of parameters in the model do not correspond with the true values.

Several methods have been investigated to deal with the impact of process-model mismatch. Garcia and Morari (1982, 1985a, 1985b) have developed design procedures that include measures of robustness for the case where linear models are used within the context of model-based controllers. When linear models are used to control highly non-linear processes, gain scheduling is a popular technique that essentially provides new model parameters based upon a local linearization. McDonald (1987) has successfully demonstrated this technique through the application of DMC for the control of a high purity distillation. The research literature abounds (Seborg *et al.*, 1986; Pollard and Brosilow, 1985; Clark *et al.*, 1987a, 1987b) with many applications of adaptive control where parameters are modified as process data has been collected. In some cases, parameters have been adapted to handle slowly changing process behaviour (Åström, 1986; Huang and Stephanopoulos, 1985).

Lee *et al.* (1989) presented a process-model mismatch compensation algorithm for model-based control. This algorithm compensated for model errors and updated the model parameters at steady state. A more practical approach has been adopted by Signal and Lee (1992) by proposing the generic model adaptive techniques.

When the system is minimum phase and in the absence of process model error, the closed loop response by GMC will follow the reference trajectory exactly. However, when the model is inaccurate, the response will deviate from the reference trajectory. The integral term in the GMC control law provides a compensation for the process-model mismatch (Lee and Sullivan, 1988). The control structure needs not to be changed if the closed loop response by this compensation is satisfactory. However, this method may not be applicable to all the systems because of the stability and the process constraints. It must be noted that mathematical conditions for accurate tracking and offset free control are much stronger than these specified reference trajectories.

When the mismatch becomes larger, the closed loop response will not be satisfactory. Also from process control point of view and for process industries, it is often required that the closed-loop response exhibit no overshoot and have suitable rise-time. These properties are ensured when a perfect process model is used or the mismatch is small, but these are not guaranteed when the mismatch and any unmeasured disturbances are large. The basic idea behind the development and application of a dynamic parameter update system described below is to cope with these difficulties and reduce the effect of larger modelling errors by regularly updating the model parameters.

7.5.2 Development of Model Parameter Update System

Signal and Lee (1992) derived an adaptive algorithm capable of adapting model parameters in a non-linear model. The algorithm was developed within a generic model control framework that reduces the effect of larger modelling errors by regularly updating the model parameters.

Consider the process described by

$$\frac{dx}{dt} = \mathbf{f}(\mathbf{x}, \mathbf{u}, \mathbf{d}, \boldsymbol{\theta}, t) \tag{7.24}$$

$$\mathbf{y} = \mathbf{g}(\mathbf{x}, \boldsymbol{\theta}) \tag{7.25}$$

where $\mathbf{x}$ is the state vector of dimension m, $\mathbf{u}$ is the input vector of dimension n, $\mathbf{d}$ is the disturbance vector of dimension l, and $\mathbf{y}$ is the output vector of dimension n, $\boldsymbol{\theta}$ is the parameter vector of dimension k and t is the time. In general $\mathbf{f}$ and $\mathbf{g}$ are vectors of non-linear functions. Equation 7.24 is the general form of the model of a non-linear process.

Consider the model

$$\frac{dx}{dt} = \hat{\mathbf{f}}(\mathbf{x}, \mathbf{u}, \mathbf{d}, \boldsymbol{\theta}, t) \tag{7.26}$$

where $\hat{\mathbf{f}}$ is an approximation to the true model function.

Signal and Lee (1992) assumed that if all the states and outputs can be measured, then it follows that

$$\dot{y} = \frac{\partial \mathbf{g}}{\partial \mathbf{x}^T} \, \hat{\mathbf{f}}(\mathbf{x},\, \mathbf{u},\, \mathbf{d},\, \theta,\, t) \tag{7.27}$$

where T is the sample time. If the setpoint of the outputs is y_{sp}, then the rate of change of y_{sp} to be proportional to the deviation from the setpoint with elimination of offset. Thus the desired closed loop trajectory is given by:

$$\dot{\mathbf{y}}_{sp} = \mathbf{k}_1 \, (\mathbf{y}_{sp} - \mathbf{y}) + \mathbf{k}_2 \int_{t_0}^{t} (\mathbf{y}_{sp} - \mathbf{y}) dt \tag{7.28}$$

where $\mathbf{k}_1$ and $\mathbf{k}_2$ are some diagonal n×n matrices. When the control is implemented digitally and the sampling time, T, is sufficiently small the integral term of Equation 7.28 can be replaced by summation.

For the process defined in Equation 7.24 to follow the reference trajectory of Equation 7.28, equate Equations 7.27 and 7.28, giving the model form of equations as:

$$\frac{\partial \mathbf{g}}{\partial \mathbf{x}^T} \, \hat{\mathbf{f}}(\mathbf{x},\, \mathbf{u},\, \mathbf{d},\, \theta,\, t) = \mathbf{k}_1 \, (\mathbf{y}_{sp} - \mathbf{y}) + \mathbf{k}_2 \int_{t_0}^{t} (\mathbf{y}_{sp} - \mathbf{y}) dt \tag{7.29}$$

The integral term in Equation 7.29 eliminates offset as discussed in details by Lee *et al.* (1989). However, a more desirable approach is to eliminate one of the sources of the offset that is process-model mismatch. This would also be useful when process parameters and/or model structures show significant time dependence.

As the effect of a change in the model parameters is related to the change in δ given by a model parameter correlation as:

$$\Delta\delta = \phi \, \Delta\theta \tag{7.30}$$

where δ is the difference between the model and the actual process and matrix ϕ is not necessarily square since the number of parameters which are being adapted can be less than or equal to the number of controlled variables. The terms $\Delta\delta$ and $\Delta\theta$ in Equation 7.30 are defined as:

$$\Delta\delta = \mathbf{L}(\mathbf{y}_{t+T} - \mathbf{y}^P_{t+T}) \tag{7.31a}$$

$$\mathbf{L}_{(i,i)} = t_i \tag{7.31b}$$

where t+T is the sampling interval and **L** is the (k x k) diagonal estimation reference trajectory matrix and determines the speed of the parameter estimation, t_i is the speed parameter and $\Delta\delta$ is the difference between the predicted and actual outputs variables over the sampling interval T. The $\Delta\theta$ is given by the equation:

$$\Delta\theta = (\phi^T \mathbf{W} \, \phi)^{-1} \phi^T \mathbf{W} \, \Delta\delta \qquad (7.32)$$

We used a weighted least squares approach to find the optimal parameter change. The diagonal weighting matrix, **W**, reflects the range of uncertainty in measuring the outputs, y and is given by:

$$\mathbf{W} = \text{diag} \left(\frac{1}{v_i^2} \, ... \, \frac{1}{v_k^2} \right) \qquad (7.33)$$

where v_i are the standard deviations of each of the measurements, y_i. The expression $(\phi^T \mathbf{W} \, \phi)$ in Equation 7.32 is non-singular. In order to calculate the new parameters we equate $\Delta\delta$ from Equation 7.31 with $\Delta\delta$ in Equation 7.32 to obtain an expression for dynamic model parameter update.

$$\theta_{t+T} = \theta_t + (\phi^T \mathbf{W} \, \phi)^{-1} \phi^T \mathbf{W} [\mathbf{L} \, (\mathbf{y}_t - \mathbf{y}^P_t)] \qquad (7.34)$$

From the above proposals, an algorithm for dynamic parameter update procedure is summarized and dynamic parameter update system for FCC reactor-regenerator is given in Figure 7.7.

7.5.2.1 Parameter Update Algorithm

Algorithm 7.1. Parameter Update

Step 1. Choose values of the GMC reference trajectory parameters for each controlled variables, calculate the matrices k_1, k_2 in Equation 7.29.
Step 2. Choose t_i, the speed parameters for the estimation of each parameter, θ_i to be updated. Calculate the matrix **L**.
Step 3. Estimate the error in the measurement and calculate the weighting matrix **W**, from Equation 7.33.
At each sampling time:
Step 4. At time t measure the process. Calculate the right hand side of Equation 7.29 and calculate the manipulated variable **u** from Equation 7.27. Predict the output at time t+T.
Step 5. At time t+T measure the output variable, $\mathbf{y}_{t+T}$ and calculate $\mathbf{L} (\mathbf{y} - \mathbf{y}^P)$ in Equation 7.31.
Step 6. Calculate ϕ, the dependence of the model on the parameters from Equation 7.30.
Step 7. Calculate the new model parameters for Equation 7.34.
Step 8. Go to Step 4.

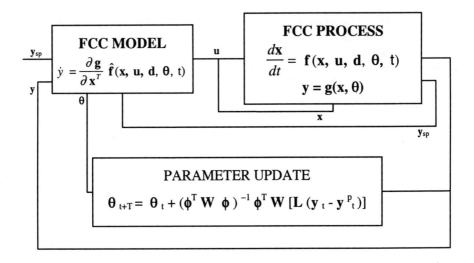

Figure 7.7. System of dynamic parameter update for FCC reactor-regenerator

7.5.2.2 Application to the FCC Process

In order to apply the dynamic parameter update program to estimate the C_{rc} and C_{cat} from Equation 7.6, a normally distributed, zero mean noise (of the standard deviation = 0.2 and 0.4 respectively) was added to each of the measurements of the controlled variables

The weighting matrix $\mathbf{W}$, in Equation 7.33, is given by:

$$W = \begin{bmatrix} 25 & 0 \\ 0 & 4 \end{bmatrix} \tag{7.35}$$

From Equations 7.6 and 7.29, the concentration of carbon on regenerated catalyst C_{rc}, was estimated with the dynamic parameter update system by using a non-square ϕ in the parameter estimation. The matrix ϕ is given by:

$$\phi = \frac{k_{cc} V_{ra} P_{ra}}{C_{cat} C_{rc}^{0.03}} \exp\left\{\frac{-Ecc}{R(T_{ra} + 460.0)}\right\} \begin{bmatrix} C_{rc}^{-0.03} \\ 0 \end{bmatrix} \tag{7.36}$$

and from Equation 7.32, the following expression is obtained:

$$(\phi^T W \phi)^{-1} \phi^T W = \frac{C_{cat} C_{rc}^{0.03} C_{rc}^{0.03}}{k_{cc} V_{ra} P_{ra} \exp\left\{\frac{-Ecc}{R(T_{ra} + 460.0)}\right\}} \begin{bmatrix} C_{rc}^{-0.06} \\ 0 \end{bmatrix} \tag{7.37}$$

In this case, the weighting matrix **W**, does not affect the parameter estimation as the system consists of five controlled variables and only one parameter is estimated in this example. Figure 7.8 shows the trajectories of the concentration of carbon on regenerated catalyst.

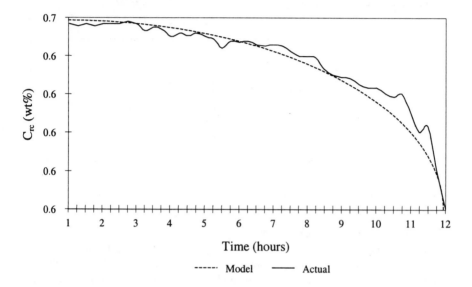

Figure 7.8. Model parameter update for FCC - estimating C_{rc}, the concentration of carbon on regenerated catalyst

The actual values were plotted against the model estimated values from the dynamic parameter update system. The sampling time T, was 2 seconds and L, the speed of parameter estimation was equal to 1. It was observed that while the model updates the parameter C_{rc} well within the region (0.65-0.70) wt. % of C_{rc}, the estimation is not good below that region. This fact may be related to structural mismatch, which can be made negligible if model parameters are regularly updated. The regenerator system of FCC is subject to measurement noise, the estimation of C_{rc} in model parameter update was good even in the presence of measurement noise. Similar approach can be applied to estimate and update the parameter C_{cat}.

It was observed in FCC application that by reducing the value of L, the speed of parameter estimation, the parameter trajectory slows down. In a closed-loop system when the model is updated regularly L requires to be tuned. The parameter update system of Signal and Lee (1992) does not address the issues related to on-line tuning of L, the speed of parameter estimation.

Also the algorithm has a limitation that the number of parameters to be updated should be less than or equal to the number of controlled variables. This restriction may not be serious on a simple system where only one parameter estimation is sufficient for one controlled variable. However, on a complex process such as FCC, where more parameters are required to be estimated on-line than the controlled variables, this restriction may cause a serious problem. Due to above mentioned limitations, the other parameters in FCC model were updated by the use of optimization software from AspenTech which contains the data reconciliation procedure in [PML] library and is suitable for real-time applications.

7.6 Model-predictive Control

7.6.1 Problem Formulation for Linear Control

The model-predictive control algorithm applied here utilizes features from established model-based control algorithms such as dynamic matrix control (DMC) and model algorithm control (MAC) as discussed in details in Chapter 2. The intent was to use features of each algorithm as they best apply to the problems associated with FCC control. The main characteristics of the control algorithm are summarized here:

1.   Discrete-step response dynamic models.
2.   Reference trajectories to define the desired closed-loop response.
3.   Quadratic objective function solved for least squared error.
4.   Iterative control calculation for constrained input operation.

These methods are well established and field proven; however, an overall control strategy requires additional programme functions which are built around the basic MPC algorithm. This is especially true for the FCC control. Additional techniques employed to improve the FCC control include:

7.6.1.1 Signal Conditioning: Control variable compensation
This technique is used in the severity control to compensate controlled variables for poor riser outlet temperature control. Oscillatory riser outlet temperature control often arises due to inadequate slide valve actuation or valve hysterisis. The compensation technique uses predictor models to offset constraint control error caused by the riser outlet temperature oscillation. The objective of the compensation is to replace the oscillatory plant response with a well-behaved response as defined by the compensator.

7.6.1.2 Prediction Trend Correction
The predicted value of controlled variables at time t= (k+1) ΔT, may be written:

$$y'_{k+1} = y'_k + H \, \Delta u_k \qquad\qquad (7.38)$$

$$\Delta u_k = u_k - u_{k-1} \qquad\qquad (7.39)$$

where y'_k is the predicted values of control variables at time k, u_k is the prediction inputs at time k, H is the step response model matrix and ΔT is the control interval.

At each control interval, the prediction calculation is corrected for the difference between the measured and predicted controlled variables. Typically the correction is applied equally to the entire prediction vector. The corrected predictions may be written as:

$$(y'_{k+1})_c = y'_{k+1} + K [y_k - y'_k] \qquad (7.40)$$

where y_k is the measured control variables at time k and **K** is the gain matrix.

The regenerator constraint variable may drift as the burning patterns in the regenerator shifts. The changes are gradual and the trends can be detected prior to the constraint violation. For example, increasing temperature trends may be detected and control actions taken before temperature constraints are violated. Here it is assumed that if a temperature has been trending upward over a period of time, it will continue to trend upward into the predicted future. The prediction correction described in Equation 7.40 is enhanced to include not only the current prediction error, but also the observed trend of past prediction errors. The trend corrected prediction is expressed as:

$$(y'_{k+1})_c = y'_{k+1} + K [y_k - y'_k] + M. E_{avg} \Gamma \qquad (7.41)$$

In Equation 7.41, $M = [0 \ 1 \ 2 \ ....(n-1)]^T$ and $E_{avg} = [e_1 \ ..... \ e_{ny}]$, e is the average prediction error, Γ is the diagonal matrix of attenuation factors, n is the prediction horizon, ny is the number of controlled variables.

Trend correction integrates an average error over the future control horizon. This correction is only added for the current control interval. This technique is effective for processes that exhibit integrating behaviour. A technique for truly integrating processes is described by Morari and Lee (1991).

7.6.1.3 Control Move Calculation
Control moves are calculated to minimize a quadratic objective function

$$J [\Delta u] = E^T QE + \Delta u \ R \ \Delta u \qquad (7.42)$$

where E is the predicted error array, **Q** is the control error weighting matrix and **R** is the input penalty weighting matrix.

The control move to minimize **J** is calculated using least squares

$$\Delta u = (A^T QA + R)^{-1} A^T QE \qquad (7.43)$$

where A is the dynamic matrix composed of step response models. The **Q** weighting matrix provides scaling for the control variable errors and allows weighting of one constraint versus another. The **R** weighting matrix penalizes excessive moves and "ringing" of the manipulated variable. Calculating a large number of future moves added significant computational loading, implementation and tuning complexity, and potential oscillatory or unstable behaviour due to ill-conditioned matrices. A more detailed discussion on this topic is presented by Marchetti et al. (1983) and Garcia et al. (1989) which suggests minimizing the number of control moves to improve control stability.

The MATLAB® toolbox on Model-predictive Control (MPC) was used for the linear control application together with the control techniques mentioned above. These techniques were incorporated in the programme and applied to the fluid catalytic cracking reactor-regenerator section. The results obtained were compared in Section 7.7.4 with the constrained non-linear multivariable control application developed for the FCC process in Section 7.4. The details of the programme are given in Appendix E.

7.6.2 Process Identification Tests

The scope of the process identification test was to capture the plant dynamics in the models to develop linear controllers such as DMC. The process identification tests were carried out by making a step change on each manipulated variable (u_1, u_2, u_3 and u_4) listed in Table 7.2. Step test for each manipulated variable lasted 2-6 hours during which time the variable was repeatedly stepped about its nominal value. In each case, the magnitude of the steps was selected to ensure that a clear response would be observed on the process. The transfer function models of the FCC process are given in Table 7.2. Similar approach was adopted by Grosdidier *et al.* (1993) during process identification test on FCC who has also discussed in details process models characterizing the FCC unit operation.

Table 7.2. FCC process models obtained by process identification tests and FCC process control objectives.

	Combustion Air Flow (u_1)	Feed Flow Rate (u_2)	Feed Preheat Temp. (u_3)	Riser Outlet Temp (u_4)	Control Object
Flue Gas Oxygen Conc.	$\dfrac{0.100(1.7s+1)e^{-2s}}{18s^2+7.0s+1}$	$\dfrac{0.90e^{-2s}}{13s^2+4.6s+1}$	$\dfrac{0.30e^{-7s}}{12s+1}$	$\dfrac{-0.080(4.8s+1)}{9.0s^2+3.0s+1}$	y_1 (set point)
Regen. Bed Temp	$\dfrac{0.03e^{-4s}}{11s^2+8.0s+1}$	$\dfrac{0.45e^{-4s}}{23s^2+8.0s+1}$	$\dfrac{0.10e^{-7s}}{10s+1}$	$\dfrac{0.080(1.7s+1)}{10s^2+7.3s+1}$	y_2 (zone limits)
Fuel Gas Flow	–	$\dfrac{0.18e^{-11s}}{40s^2+8.5s+1}$	–	$\dfrac{0.37(16s+1)}{50s^2+20s+1}$	y_3 (max. zone limit)
WG Output	–	$\dfrac{0.30e^{-11s}}{16s^2+7.0s+1}$	–	$\dfrac{0.60}{3.0s+1}$	y_4 (max. zone limit)
Riser Out Temp Output	–	$\dfrac{0.72e^{-s}}{2.5s+1}$	$\dfrac{-0.8e^{-10s}}{6.0s+1}$	$\dfrac{1.5}{2.0s+1}$	y_5 (max. zone limit)

In the following subsection, some of the important process models obtained by the process identification tests are discussed.

7.6.2.1 Combustion-air-flow Models

The Combustion air flow models are shown in the first column of Table 7.2. In all the FCC units operating in full burn mode, coke combustion is essentially complete and increasing the flow of combustion air to the regenerator can only raise the flue gas oxygen concentration. This effect explains the positive gain between air (u_1) and oxygen (y_1) in Table 7.2. The last three controlled variables did not respond to changes in combustion air flow, therefore no transfer functions appear between air and the last three controlled variables in the first column of Table 7.2.

7.6.2.2 Feed-flow-rate Models

Feed Flow Rate models are shown in the second column of Table 7.2. Increasing the feed flow rate to the riser increases the total amount of coke and vapours produced by the unit. The combustion of the additional coke consumes some of the excess oxygen in the regenerator and therefore, lowers the flue gas oxygen concentration. The increase in the vapour rate raises the load on the wet gas compressor and increases the fuel gas make. The additional load on the wet gas compressor is met by the suction pressure controller, which admits more steam to the compressor's turbine. These effects explain the sign of the gains in the transfer functions between feed and oxygen, fuel gas and WGC suction pressure controller output. The response of bed temperature to an increase in the feed rate resulted in increasing the bed temperature due to combustion of the additional coke produced. When a unit processes light feedstocks for which the coke yield is low, it is common to see a negative gain between feed flow rate and regenerator temperature as shown in Grosdidier (1990). In Table 7.2, the gain is positive because more feed is charged to the riser, the cooling effect of increased catalyst circulation is largely compensated by the heating effect of greater coke combustion.

7.6.2.3 Feed-preheat-temperature Models

Models relating the feed preheat temperature to process outputs are shown in the third column of Table 7.2. Changes in feed preheat temperature, like changes in charge rate, introduce a load on the riser outlet temperature controller. At constant cracking temperature, coke yield decreases with decreasing catalyst to oil ratio. This effect assigns a positive gain to the transfer function between the feed temperature and the flue gas oxygen concentration and has a cooling effect on the bed. Very small gain on bed temperature response was observed after stepping the feed preheat temperature. It should be noted that on other FCC units, the bed temperature may respond with either a positive or negative gain depending on the unit's operating conditions and the nature of its feed. Finally, neither fuel gas nor WGC suction pressure output showed any significant response to changes in feed temperature.

7.6.2.4 Riser-outlet-temperature Models

The effects of changing riser outlet temperature setpoint on the process are straightforward to understand. Raising riser temperature increases the catalyst circulation and both the dry gas and coke yields. These effects are sufficient to explain the sign of the steady-state gains in all the transfer functions in the fourth column of Table 7.2.

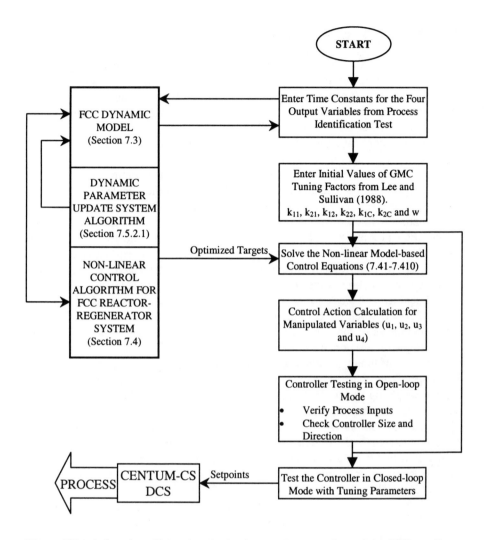

Figure 7.9a. A flowchart illustrating the implementation procedure of the FCC non-linear model-based multivariable control

7.7 Real-time Implementation

There are three important steps to implement the multivariable control applications in real-time. The first step is to follow the exact procedure and sequence of implementation. Figure 7.9a illustrates the implementation procedure of the FCC non-linear model-based multivariable controller. The second step is to configure the controller interface to distributed control system and finally the on-line controller tuning to meet the specified targets without process deviation and to gain the confidence of the operators.

7.7.1 Non-linear controller tuning

The time constants of the temperature controllers residing in a distributed control system (DCS) were estimated from open-loop tests. The time to steady state was found to be about 80 minutes. The sampling frequency was taken once per minute. The desired trajectories of the process outputs were determined using the tuning rules defined by Lee and Sullivan (1988).

The control application was first turned on and tested in an open-loop mode (*i.e.*, the outputs were suppressed). Open-loop testing includes verifying process inputs and checking controller move sizes and directions. Next, the controls were tested in closed-loop mode using tuning constants gathered from the simulation model and step response test. Closed-loop testing was performed within a narrow operating range until confidence was gained. It was also noted that the form of the optimization problem is well structured since a slack variable is added to each control law equation to ensure that a solution to the set of equations does in fact exist. It is important to note that if the control is implemented at a reasonable frequency, the solution of the NLP is very fast (3 to 4 iterations), since the current control settings and slack variables provide a good initial estimate of the solution vector. Further, when the control system fails or turns itself off, the basic control system continues to function. In effect, the operator does not have to do anything special to put the reactor-regenerator on computer or to take it off.

Two types of tuning parameters were used; the K_{1C} and K_{2C} (Equations 7.16 and 7.17), specifying the maximum speed of approach towards the constraint bounds, and the weighting factors W (Equation 7.13), reflecting the relative importance of the outputs and the constraints. The tuning parameters K_{1C} and K_{2C} can be compared with the *move suppression factor*, a term used in linear multivariable controller design such as dynamic matrix control (DMC). Changing the move suppression factor on a manipulated variable causes the controller to change the balance between movement of that manipulated variable, and errors in the dependent variables.

The role of the weighting factors, W, is similar to the *equal concern factor* (controlled variables priority) in a linear multivariable control framework such as (DMC). In the regenerator loading controller tuning analysis, the weighting factor for the regenerator temperature (controlled variable) was assigned a value of $1°C$. It was observed that $1°C$ on the regenerator bed temperature was as important as 2 tons/hr air flow. If the controller was unable to reach all the setpoints or limits, it would prioritize the errors in the ratio $1°C=2$ tons/hr.

7.7.2 Controller Interface to DCS System

The non-linear multivariable controller was interfaced to Yokogawa Centum-CS DCS system as shown in Figure 7.9b. The system consists of a Unix-Based engineering workstation (EWS) which was used for M-series signals, collecting and converting tag data for the development of process models. The PC platform was used to create on-line tuning file of the controller. Non-linear controller resides in advanced control station (ACS) and using the standard ACS load utility of the

EWS, the controller were loaded off-line to ACS of Centum-CS. Corresponding to each model built, the non-linear controller configuration were defined. After compiling the configuration file, the controllers were loaded on-line. Non-linear controllers displays were built in information and command station (ICS), which is used only by operators to monitor and run the controllers.

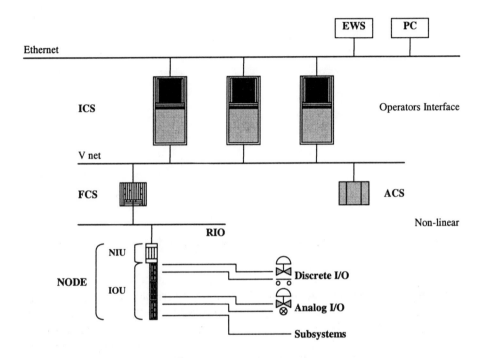

Figure 7.9b. Non-linear controller interface to Centum-CS DCS system

7.7.3 Results and Discussion

The response of the process models obtained through the process identification tests are given in the Figures 7.10 to 7.13 and the Table 7.2 gives the transfer functions of the FCC unit. This table also defines the process control objectives of each variable.

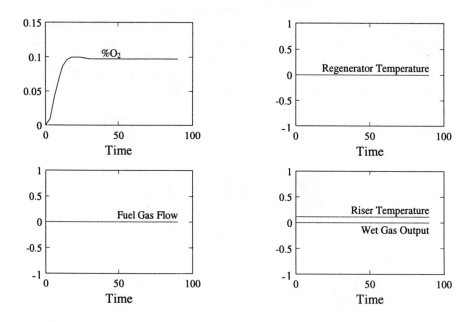

Figure 7.10. Responses of the controlled variables to a step change in the combustion air flow (u_1)

Figure 7.10 shows the responses of the controlled variables to a step change in the combustion air flow (u_1). This model is explained in Section 7.6.2. Figure 7.11 gives the responses of the controlled variables to a step change in the feed flow rate (u_2) and the model is explained in Section 7.6.2. The responses of the controlled variables to a step change in the feed preheat temperature (u_3) is shown in Figure 7.12 and the model is explained in Section 7.6.2.

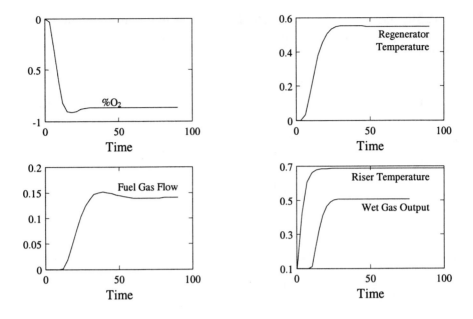

Figure 7.11. Responses of the controlled variables to a step change in the feed flow rate (u_2)

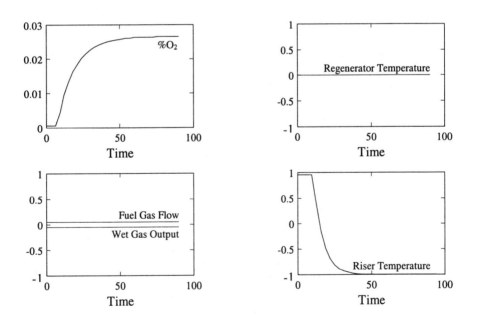

Figure 7.12. Responses of the controlled variables to a step change in the feed preheat temperature (u_3)

Finally, the responses of the controlled variables to a step change in the riser outlet temperature ($\mathbf{u}_4$) are given in Figure 7.13 and the model is discussed in Section 7.6.2.

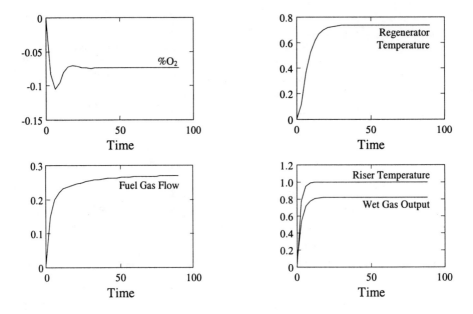

Figure 7.13. Responses of the controlled variables to a step change in the riser outlet temperature ($\mathbf{u}_4$)

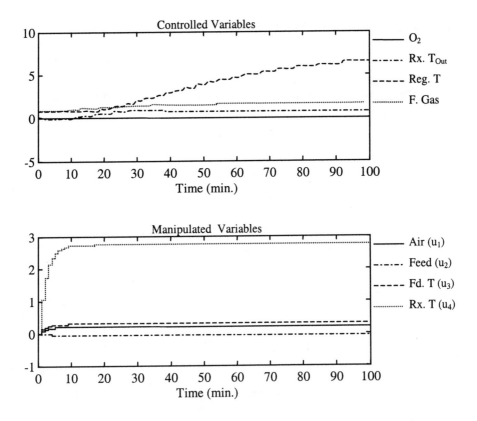

Figure 7.14. Controlled variables response for unconstrained case

The controller performance was checked for unconstrained case with minimum manipulated variable move $|\Delta MV| = 2.75$. The controlled variables response is shown in Figure 7.14. These figure shows that while O_2 concentration and riser outlet temperature controller output are least affected by the manipulated variable move, the regenerator temperature rises significantly.

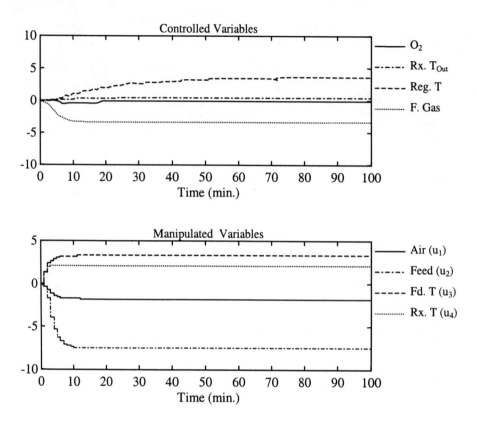

Figure 7.15. Constrained case: process variables constrained to ± 5

Figure 7.15 shows the response of the controlled variables when process variables are constrained to ±5. The O_2 and riser outlet temperature controller output are still not affected by the manipulated variables move, min. $|\Delta MV| =$ 8.50. The regenerator temperature and fuel gas values remain within the constraint limits. This demonstrates the constraint handling capabilities of the multivariable control.

Figure 7.17 shows the response of the controlled variables when both the process variables and manipulated variables are constrained. The minimum manipulated variable move, min. $|\Delta MV| = 6.23$. A reduction in the feed has resulted in a reduction of fuel gas, and therefore a lower conversion and less effect on O_2 concentration and riser temperature controller output result. This fact may also be observed in the operation of FCC units.

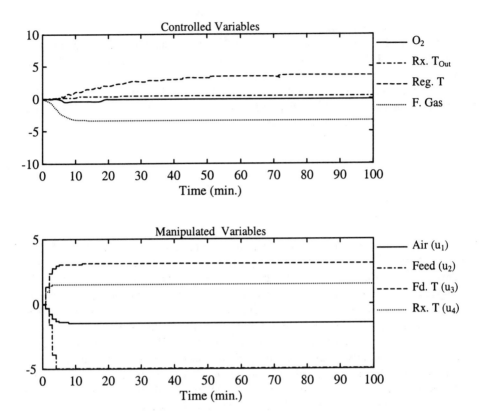

Figure 7.16. Results for when controlled and manipulated variables are constrained

7.7.4 Comparison of Non-linear Control with DMC

In this section non-linear controller performance is compared with the dynamic matrix control by simulation results and with the results obtained from real-time applications.

Table 7.3. Setpoint, operating limits and GMC parameters

Output	Setpoint/ Limits	GMC Parameters	
		ξ	τ (min)
O_2 conc.	1.0 %	5	70
Reg.T	714-715 °C	5	70
RxT_{out}	80%	5	70
F.Gas	15 tons/hr	5	70

7.7.4.1 Simulation Results

In the following figures, linear control performance was compared with the non-linear control. Table 7.3 shows the setpoint and controlled variables operating limits and GMC parameters for non-linear controller.

For each output variable in Table 7.3, the GMC controller parameters K_{1i} and K_{2i} (the ith diagonal elements of K_{11} and K_{21}) from Equations 7.14 and 7.15 were calculated by the following relationships:

$$K_{1i} = \frac{2\xi_i}{\tau_i} \tag{7.7.1}$$

$$K_{2i} = \frac{1}{\tau_i^2} \tag{7.7.2}$$

τ_i was estimated from the process identification tests (plant test runs) and ξ_i was selected from the GMC performance specification curve (Lee and Sullivan,1988). This pair of parameters has explicit effects on the closed-loop response of the output variable. The parameter ξ_i determines the shape of the closed-loop response while τ_i determines the speed of the response (large τ_i implies slow response). The GMC design parameters were selected as discussed in Section 6.3.2.

Figure 7.17 compares the linear control with non-linear control in terms of decoupling capability of the controllers. Since the FCC process is non-linear and time-variant, decoupling capability of the non-linear controller provides better performance as compared to linear controller. It can be observed from Figure 7.17 that the non-linear controller displays a good decoupling capability.

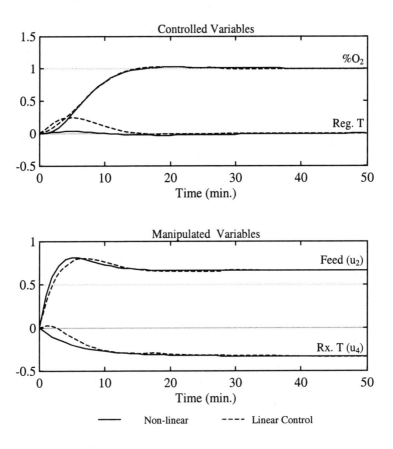

Figure 7.17. Comparison of the decoupling capability of linear and non-linear controllers

Consider the response of O_2 concentration in Figure 7.13. The Oxygen concentration shows the presence of nonminimum-phase characteristics due to the right-half plane (RHP) zeros and poses difficult control problems in linear systems. For a linear system, the zeroes are the poles of the process inverse and are nonminimum phase if the process inverse is unstable. For a non-linear system, there is no explicit definition for process zeroes and poles. Figure 7.18 compares the performance of linear and non-linear controllers in controlling the process with inverse response. This figure shows that tuning the move suppression factor in linear control to avoid inverting RHP zero is difficult. Non-linear control in that regards gives a better performance.

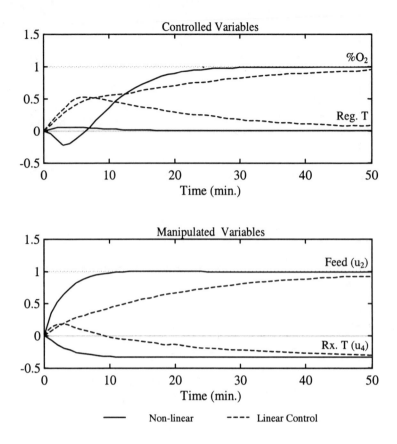

Figure 7.18. Comparison of linear and non-linear controllers in controlling the process with inverse response

Figure 7.19 compares the effect of tuning on the performance of linear and non-linear controllers. This figure shows that under linear control small changes in manipulated variable have significant impact on O_2 concentration and regenerator temperature and the control action is much more aggressive and instability can be observed for a large number of manipulated variable move. However, the non-linear control performance under the same conditions is far better and smooth. This also demonstrates that tuning the non-linear controller is much easier than the linear controller.

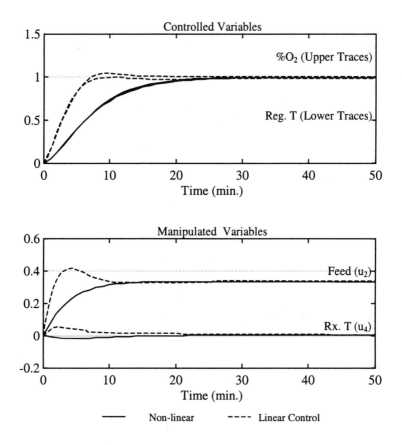

Figure 7.19. Effect of tuning parameters on the performance of linear and non-linear controllers

7. 8 Plant Results

Figures 7.20 and 7.21 compare the non-linear controller performance with the DMC in real-time application on the FCC process. Figure 7.20 shows the regenerator bed temperature control trends. The regenerator bed temperature is required to be controlled between the limits (714-715) °C as shown in Table 7.3. The DMC controller shows that the temperature control sometimes exceeds the desired range (deviation = 1°C), however, the non-linear controller keeps the temperature well within the range (deviation = < 0.5°C). Both the controllers perform well, however, under non-linear control, the regenerator bed temperature response is much more stable. This may be explained as follows, since the regenerator bed temperature is very much affected by the coke formation on the regenerator bed catalyst, the two parameters defined in coke formation rate Equation 7.6 are updated regularly by the parameter update system, which improves the performance of the non-linear controller over DMC.

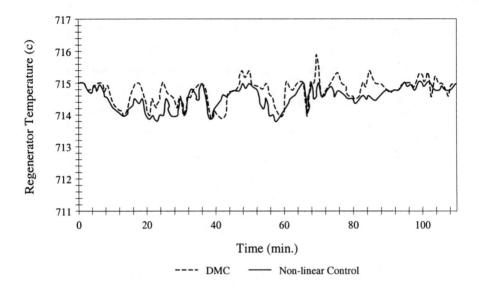

Figure 7.20. Comparison of non-linear control performance with DMC: real-time trends of regenerator bed temperature control

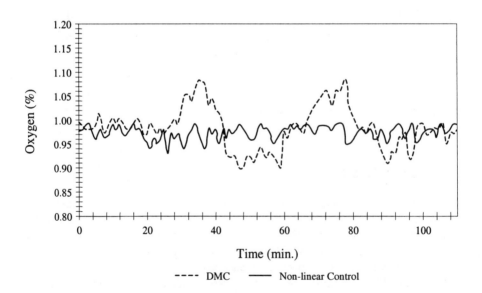

Figure 7.21. Comparison of non-linear control performance with DMC: real-time trends of flue gas oxygen concentration control

Figure 7.21 shows the oxygen concentration control trends for both the controllers in real-time application. The primary control objective is to minimize the oxygen concentration in flue gas analysis to a setpoint value of 1% vol. The

DMC controller keeps the oxygen concentration between the limits (0.9-1.1)%vol. (deviation of 0.2% vol. in oxygen concentration). Under the non-linear controller, the response is far better as the deviation from the setpoint is very small (0.05% vol.).

7.9 Conclusions

The fluid catalytic cracking is the most complex and challenging operating process in the modern refinery. The success of any advanced control application (linear or non-linear) on this unit depends on the ability to deal with the unit operating constraints and to address its economic objectives. Economic objectives are defined by the economic optimal point at which to operate the process. A constrained non-linear optimization strategy for handling the constraints has been developed and applied to the fluid catalytic cracking reactor-regenerator section. A dynamic parameter update model was developed and used to provide target values for an important controlled variable to compensate for the process/model mismatch.

The non-linear control strategy presented in this chapter translates operating constraints into setpoint and zone limit objectives for the controlled variables of the process. Because it is more important to satisfy operating constraints than to meet economic objectives, the non-linear controller moves the manipulated variables in such a way that all the controlled variables remain at their setpoints or within their zone limits. In this manner, the process remains as close as possible to its optimal operating point while at the same time ensuring that no operating constraint is violated.

A simplified form of non-linear dynamic model from (Denn, 1986) was used as a reference for the design of the control algorithm. The model results were compared with the plant data, for open loop changes on the air flow. The control problem of the FCC reactor-regenerator section was analyzed with an extension of the conventional model-predictive control (MPC) algorithm using MATLAB® control features. The results of the linear MPC were compared with the non-linear model-based multivariable control. It has been shown that non-linear multivariable control has provided better decoupling ability and control of the process with inverse response as compared to traditional linear model-based control. It has also demonstrated that tuning the non-linear controller is much easier than the linear controller. The same model is used for optimization and control, minimizing the modelling errors due to process/model mismatch.

The main contribution of this work is to incorporate non-linear process model directly into non-linear constrained multivariable control strategies and the application of non-linear optimization algorithm to a highly non-linear fluid catalytic cracking process. In addition, a dynamic model parameter update algorithm was developed and used to update the important parameters in the FCC process model. The MATLAB® toolbox on Model-predictive Control (MPC) was used for the linear control application incorporating the control techniques in the program such as control variable compensation and prediction trend correction as described in Section 7.6. For the non-linear system, constrained non-linear optimization problem was applied to the FCC reactor-regenerator section and the solution was obtained by using the Optimization toolbox of MATLAB®. The **CONSTR.M** file was used to find the constrained minimum of the objective function by using the sequential quadratic program (SQP) method. From the Non-linear Control Design (NCD) toolbox, a non-linear optimization file **NLINOPT.M** was used to run the optimization algorithm.

The process model played a vital role in the optimization strategy. It was used to make the outputs follow their trajectories in Equations 7.14 and 7.15 and to avoid the constraint violations of regenerated bed temperature in Equations 7.16 and 7.17. It shows that the accuracy of the model is very important for the success of this strategy. Although, the integral term in GMC control law provides the compensation for the process-model mismatch, implementing a dynamic model parameter update algorithm given in Section 7.5, accounted for model inaccuracy problem by regularly updating the two important parameters in Equation 7.6. Signal and Lee (1992) also adopted this approach by proposing the generic model adaptive techniques.

In real-time application, the constrained non-linear optimization strategies were tested and compared to DMC. The main advantage of non-linear control approach, besides the improved control performance, is that a single-step control law resulted in a smaller dimensional programme as compared to other available methods and the time spent to do the plant testing to obtain the process models was reduced significantly. The other advantage of this approach is that since the same model was used for optimization and control, it minimizes the maintenance and process re-identification efforts which is often required for linear controller as the operating conditions keep changing all the time.

CHAPTER 8
CONCLUSIONS AND RECOMMENDATIONS

8.1 Conclusions

8.1.1 Summary of Results Achieved

The specific objectives of this study, stated in Section 1.3, have been achieved. The overall scope of this study described in Section 1.4 has also been completed. The non-linear model-based multivariable control algorithms and constrained non-linear optimization strategies were developed and an integrated approach of the control strategy was proposed and applied to various refinery processes. This integrated approach entailed incorporating a process model, an inferential model and multivariable control algorithm and constrained non-linear optimization strategies in one framework. The scope of work summarized in Section 1.4 has been successfully completed. The main features of the completed work are as follows:

1. Extended the non-linear model-based control structure of Lee and Sullivan (1988) to permit the use of inferential models in non-linear multivariable control applications.

2. Developed a wide range of inferential models such as the iso-pentane (iC$_5$) and RVP for debutanizer; naphtha final boiling point and kerosene flash point for crude fractionator and octane number for catalytic reforming process. Implemented these models in real-time and integrated with non-linear multivariable control applications to form a closed-loop quality control.

3. Designed and developed non-linear control strategies for industrial debutanizer using steady state model with approximate dynamics and implemented the control strategies in real time.

4. Developed a complex non-linear multivariable control structure by formulating the constrained non-linear optimization method that optimizes the performance objectives for crude distillation process subject to constraints. Shell heavy oil fractionator was selected for this purpose, the process dynamic model was built by process identification tests and the model uncertainty was considered

explicitly in non-linear controller design. The non-linear control strategies were tested on a 120,000 barrels per day crude-fractionator.

5. Developed a constrained non-linear optimization algorithm for the reactor section of semi-regenerative catalytic reforming process. Also simulated a dynamic model of the catalytic reforming process and developed an inferential model to predict the reformate octane. Integrated all these models in non-linear model-based control structure and implemented control strategies in real-time.

6. Extended the application of constrained non-linear multivariable control and optimization to more complex process such as the fluid catalytic cracking (FCC) and developed the dynamic parameter update algorithm for the FCC to reduce the effect of larger modelling errors by regularly updating the selected model parameters. Developed a simplified dynamic model of the process and tested the model in real-time to provide the initial targets for the non-linear controller.

8.1.2 Outline of the Major Contributions of this Research

The current interest of the processing industry and the latest trends in academic research in model-based control have been highlighted and many industrial applications of model-predictive control have been reported. The need for non-linear model-based control is emphasized, as model-based control algorithms based on linear model can not handle non-linear systems very well, especially when the non-linearity becomes strong. Stability and robustness issues related to both the linear and non-linear control systems are also discussed. The relations between generic model control (GMC) and internal model control (IMC) and model-predictive control (MPC) have been investigated. It was found that for linear systems, if the model is perfect, GMC is equivalent to IMC plus two special filters. These two filters lead to the direct application of state space process model rather than transfer functions. *This also implies that GMC is applicable to both linear and non-linear systems.* The discrete GMC control law was derived for a discrete process model. It was concluded that different schemes of differential and integral approximation would result in different forms of the control law. The major difference between GMC and MPC is in the usage of past process measurements and past control action, which is partly due to the different output reference trajectories and different forms of the process model.

A wide variety of inferential models were developed starting from naphtha final boiling point and kerosene flash point of crude distillation; iso-pentane and RVP of stabilizer and reformate octane number of the catalytic reforming process. These inferential models have been shown to perform well in on-line applications on heavy oil fractionator, debutanizer and catalytic reforming process. The on-line applications interfacing with non-linear multivariable control formed a closed-loop quality control system, improving the dynamic performance of multivariable control system by minimizing the long time delays caused by the analyzers. The application helped process operators achieve quality targets, resulting in less quality giveaway and more optimal yield. Operator's acceptance of these models was very high as these models have eliminated the long waiting hours for the laboratory results as the prediction of these models was within the repeatability of laboratory results.

The non-linear model-based control structure proposed by Lee and Sullivan (1988) was extended to permit the use of inferential models in non-linear control strategy, which was applied to a debutanizer. This strategy has been shown to provide improved control performance of the top product and the bottom stream qualities using a steady-state process model with approximate dynamics. The control performance was much better than that achieved by the traditional PID-type control system. The use of inferential models as controlled variables in the non-linear model-based control has been very rewarding both in terms of reducing the long-time delays and enhancing the potential of the non-linear control to keep the product qualities close to their setpoints. These models also assisted the operators to drive the process closer to the optimum. Other interesting feature of this non-linear multivariable control application was that it was implemented on a hardware system without distributed control functionality. This suggests that a distributed control system is not a prerequisite hardware for multivariable control implementation.

The heavy oil fractionator control problem proposed by Prett and Morari (1987) was solved by incorporating the dynamic process model obtained from the open-loop tests in non-linear constrained optimization strategies. In addition, an inferential model for the top end point (TEP) was developed and integrated with the process model showing a better response in case of analyzer failure. The results of the linear MPC were compared with the non-linear model-based multivariable control. It has been shown that non-linear multivariable control provided better overall disturbance rejection for the end points. The simulation studies conducted show that both controllers exhibit good qualitative robust performance characteristics. However, non-linear controller, due to its flexibility in dealing with constraints is well suited for solving this problem. In real-time application, the performance of the non-linear controller was compared with the linear controller. It was demonstrated that the non-linear controllers have better ability to keep the end point targets within the lower and upper constraint and to control the product qualities to specification for a wide range of operating condition.

A constrained non-linear optimization strategy for handling the constraints has been developed and applied in real-time to the catalytic reforming reactor section. An inferential model predicting the octane number was also developed and integrated with multivariable controller forming a closed loop WAIT/octane control. A dynamic model of the catalytic reforming process was developed and used to provide target values for the reactor inlet temperatures. It has been shown that non-linear multivariable control provided better disturbance rejection compared to the traditional linear model-based control. The same model was used for optimization and control, minimizing the modelling errors due to process/model mismatch. The main contribution of this work was to combine non-linear multivariable controller with non-linear constrained optimization models and its application to a highly non-linear catalytic reforming process. Through the optimization formulation, it has been shown that the merit of this technique is the degree of flexibility in establishing the proper balance between the violation of the constraint variables and the deterioration of the control performance coupled with the ability to pre-define the response trajectories for both the controls and constraints.

A non-linear constrained optimization strategy was proposed and applied to fluid catalytic cracking reactor-regenerator section. A dynamic parameter update algorithm was developed and incorporated in the optimization strategy to reduce the effect of larger modelling errors by regularly updating the model parameters. This algorithm is capable of adapting the model parameters in a non-linear model.

The results of the linear MPC were compared with the non-linear model-based multivariable control. It has been shown that non-linear multivariable control provided better decoupling ability and control of the process with inverse response as compared to the traditional linear model-based control. It was also demonstrated that tuning the non-linear controller is much easier than the linear controller. In the optimization strategy, maximum rates of approach towards the constraint bounds were specified. Slack variables were introduced both for the output performance and for the constraints. These slack variables are measurements of the performance degradation and the potential constraint violation.

The process model played a vital role in this strategy. It was used to make the outputs follow their trajectories and to avoid the constraint violations of regenerated bed temperature. It shows that the accuracy of the model is very important for the success of this strategy. As there is always a Process/Model mismatch, implementing a dynamic model parameter update algorithm given in Section 7.5, accounted for model inaccuracy problem.

8.2 Recommendations

Several interesting research topics have arisen during the course of this study and these topics are recommended for future investigation and development.

8.2.1 Embedded Optimization for Non-linear Control

The constrained non-linear optimization problem developed in Chapter 5 for Shell heavy oil fractionator points to the deficiencies of optimizing multiple performance criteria via a lumped objective function approach. A more appropriate course of action is to handle the multiobjective optimization criteria at different levels within a single mathematical framework. A possible future research direction would be to develop a non-linear multivariable controller, which can realize maximum benefits from all objectives while simultaneously satisfying the constraints. A possible method for obtaining this type of controller is to use embedded optimization. With this approach, optimization of the performance specifications would be posed explicitly as a constraint in a higher level "economic" optimization. More specifically, the solution of one optimization problem would be formulated as a constraint within a second optimization problem.

8.2.2 Non-linear Nonminimum Phase Systems

The nonminimum phase behaviour was discussed in Chapter 7 for the catalytic cracking process where the oxygen concentration showed the presence of nonminimum-phase characteristics due to the right-half plane (RHP) zeros and posed a difficult control problem. For a linear system, the zeroes are the poles of the process inverse and are nonminimum phase if the process inverse is unstable.

For a non-linear system, there is no explicit definition for the process zeroes and poles. Although nonminimum phase response is exhibited only by a small number of processing units (Stephanopoulos, 1984), it does cause extra control difficulties (Shinskey, 1988). For linearizable non-linear systems, nonminimum phase systems are defined as those systems whose zero dynamics are not asymptotically stable (Henson and Seborg, 1990). From the viewpoint of a differential geometric approach (DGA), Henson and Seborg (1990) claimed that for nonminimum phase systems the full state linearization technique should be used rather than the input-output linearization technique. This is because the zero dynamics are inherent properties of the systems (under linearization). By extending the concept of "first-order all pass" from the linear area to non-linear area, Kravaris and Daoutidis (1990) developed a method to determine a static state-feedback control law for a special kind of SISO nonminimum phase non-linear system which is control-linear and has only two state variables. This control law provides ISE-optimal response to step change in the setpoint. To et al. (1995) established the equivalence of GMC with the input-output linearization. Further effort is needed in the area of controlling non-linear nonminimum phase systems.

8.2.3 Robust Stability and Performance of Non-linear Systems

The development of a workable framework for the analysis of robust stability and performance of constrained, non-linear controller is still an open question. This applies not only to the optimization based approaches but to many of the companion approaches as well (global linearization, reference system synthesis, etc.). The analysis of robust non-linear control systems remains a difficult problem; much work is required to bring it to the level of robust control for linear systems without constraints (Morari and Zafiriou, 1989). The main limitation rests with the difficulty of finding <u>global stability</u> properties for non-linear systems, which are obtained, automatically in the linear case. Often, even determining local stability properties is a difficult task as well (Economou, 1986). Thus, further research should concentrate on developing and extending tools for stability and performance analysis. Robust stability analysis of GMC was discussed in Chapter 2 under the condition that the explicit control law was available. Further studies are also needed for the implicit control law, i.e., the optimization form of the control law.

8.2.4 On-line Parameter Estimation (Model Adaptation)

As proposed by Signal and Lee (1992), a dynamic parameter update system was developed in Chapter 7 for FCC unit, which was capable of adapting model parameters in a non-linear model. This algorithm has a limitation that the number of parameters to be updated should be less than or equal to the number of controlled variables. On a complex process such as FCC, where more parameters are required to be estimated on-line than the controlled variables, this restriction may cause some problems. A dynamic parameter system is required to be developed to overcome this restriction. On line estimation of the parameters is an important consideration to improve the accuracy of the process model. The change of the model parameters and /or the model structure usually implies the necessity of changing the parameters of the controller for high control performance. Investigating an appropriate frequency of the parameter estimation (model adaptation) and its interaction with the control algorithm is an important area for further research. While this will not necessarily lead directly to the refinement and development of control algorithms, such an effort will offer a powerful framework for evaluation of current and future non-linear algorithms (Biegler and Rawlings, 1991).

APPENDIX A

A.1 Program for Pressure-compensated Temperature

```
-- CL PROGRAM FOR PRESS COMPENSATED TEMPERATURE
-- IMPLEMENTED IN HONEYWELL TDC3000 DCS AM SYSTEM

T_COMP.CL
BLOCK VLR_CALC (POINT TCOMP002; AT PV_ALG)
EXTERNAL TI002, PC002
LOCAL T_COMP
LOCAL CC = -5000    ----CLAUSIUS CONSTANT

--- PRESS COMPENSATED CALCULATIONS

SET T_COMP = (CC/
(CC/(TI002.PV+273)+LN(2.3/(PC002.PV+1)))) -273
-- TI002 (IN KELVIN)
-- PC002 (IN ABSOLUTE UNITS)
CALL ALLOW BAD (PVCALC, T_COMP)
SET PVAUTOST = (WHEN BADVAL (T_COMP):BAD; WHEN OTHERS:
NORMAL)
END VLR_CALC
```

A.2 Program for Naphtha-final-boiling-point Inferential Model

```
-- CL PROGRAM FOR NAPHTHA FBP INFERENTIAL MODEL
--This program can be used for both linear and non-
linear systems

NFBP.CL
BLOCK NFBP (POINT NFBP002; AT GENERAL)
EXTERNAL FR1002, FC2002, TCOMP002, FBPQL002, FBPLB002,
CAL, FILTER
EXTERNAL BIAS, LABOLD
LOCAL FBP

IF LABOLD.PV = FBPLB002.PV THEN GOTO A
```

```
SET BIAS.PV = FILTER.PV * BIAS.PV + (1-FILTER.PV) *
(FBPLB002.PV-FBPQL002.PV)
SET LABOLD.PV = FBPLB002.PV

-- LINEAR CORRELATION

A: SET FBP = TCOMP002.PV + LN (FR1002.PV/FC2002.PV) +
CAL.PV + BIAS.PV

-- NON-LINEAR CORRELATION

A: SET FBP = TCOMP002.PV +  (FR1002.PV/FC2002.PV) +
CAL.PV + BIAS.PV
CALL ALLOW_BAD (FBPQL002.PV, FBP)
END NFBP
```

A.3 Theory Underlying Pressure-compensated Temperature

An equivalent expression for Tcomp. in Equation 3.8 is given by the following expression:

$$\text{Tcomp} = [[CC/(CC/(\text{Top Temp}+273)+LN(2.3/(\text{Press}+1)))] - 273 \qquad (A.1)$$

where Top Temp is in Kelvin and Press. Is given in absolute units. CC is the Clausius-Clapeyron parameter. The expression A.1 was used in real-time application. The parameter CC or calibration constant is very important. It varies with the application. In general, in crude distillation applications for light ends such as C_3, C_4, C_5 or LPG its value is lower than the heavy ends such as naphtha, kerosene or light diesel oil. Therefore, the correct value of this parameter in real-time is very important for any inferential model to predict within the operating range. Following is the method to derive this parameter.

Consider the Clausius-Clapeyron equation of the form:

$$\left(\frac{\delta P}{\delta T}\right)_{eq} = \frac{\Delta h_{vap}}{\Delta v T} \qquad (A.2)$$

Δhvap is the heat of vaporization at temperature T and Δv is the difference in volume of the substance in the vapour and liquid phases.
Suppose the vapour phase behaves like an ideal gas then: $V_{vap} > V_{liq}$ and

$$V_{vap} = RT/P \qquad (A.3)$$

Equations A.1 and A.2 give the following relation:

$$\frac{\delta P}{P} = \frac{\Delta hvap}{R} \frac{\delta T}{T^2} \qquad (A.4)$$

From Equation A.4, it follows that:

$$P\left(\frac{\delta T}{\delta P}\right)_{eq} = \frac{RT^2}{\Delta hvap} \qquad (A.5)$$

Integrating Equation A.5; assuming that $RT^2/\Delta hvap$ is constant and denoted as CC, the following expression is obtained:

$$\text{Ln } P_2/P_1 = \Delta hvap/RT^2 (T_2 - T_1) \qquad (A.6)$$

$\text{Ln } P_2/P_1$ can be expanded into a Taylor series as:

$$\text{Ln } x = x-1/x + \frac{1}{2} (x-1/x)^2 + 1/3 (x-1/x)^3 + \ldots\ldots\ldots, \text{ where } x = P_2/P_1$$

Using only the first term of Taylor series; Equation A.6 takes the form:

$$1 - P_1/P_2 = \Delta hvap/RT^2(T_2 - T_1) \qquad (A.7)$$

$$RT^2/\Delta hvap = CC = (T_2 - T_1)/ 1 - P_1/P_2 \qquad (A.8)$$

Equation A.8 is the final form of expression used to determine the parameter CC for various applications.

APPENDIX B

B.1 S-B GMC Controller Implementation

The following is the set of equations to implement S-B GMC Controllers:

1. Calculate the final values of the manipulated variables (x, y) using the GMC controller equations:

$$x_{ss} = x + k_{1,1}T_x(x_{sp} - x) + k_{2,1}T_x\int_0^t (x_{sp} - x)dt \qquad (B.1)$$

$$y_{ss} = y + k_{1,2}T_y(y_{sp} - y) + k_{2,2}T_y\int_0^t (y_{sp} - y)dt \qquad (B.2)$$

$$w_{ss} = w + k_{1,3}T_w(w_{sp} - w) + k_{2,3}T_w\int_0^t (w_{sp} - w)dt \qquad (B.3)$$

The terms involved in these equations have been discussed in Chapter 4.

2. Calculate external material balance required to get y_{ss} and x_{ss}, given filtered values of W and z:

$$D = w\,(z - x_{ss})/(\,y_{ss} - x_{ss}) \qquad (B.4)$$

$$RB_{Flow} = W - D \qquad (B.5)$$

Since the feed composition does not change, $z = 0$; which makes the Equation B.4 to take the following form:

$$D = x_{ss}/(x_{ss} - y_{ss}) \qquad (B.6)$$

3. Calculate feed temperature (T_f), top temperature (T_t) and bottom stage temperature (T_b):

$$xk_{1,1}(T_b, P) + (1-x)\,k_{2,1}(T_b, P) = 1 \qquad (B.7)$$

$$yk_{1,2} (T_t, P) + (1-y) K_{2,2} (T_t, P) = 1 \qquad (B.8)$$

$$k_{2,3} (T_f, P) = 1 \qquad (B.9)$$

4. Calculate recovery (f), to get the desired product from the equation:

$$f = RB_{Flow} (1-x_{ss})/W \qquad (B.10)$$

5. Solve the remaining equations for reflux ratio R and feed stage temperature T_f

$$R_{Flow} = R D \qquad (B.11)$$

where R, the reflux ratio and is given by R_{Flow} /D

$$R_{Flow} \ P = W + R_{Flow} \qquad (B.12)$$

$$RB_{Flow} = R_{Flow} P - B \qquad (B.13)$$

$$RV_{Flow} = RB_{Flow} \qquad (B.14)$$

where RV_{Flow} is the vapour flow rate above the feed stage.

6. Calculate average stripping and rectification temperatures, T_n and T_m:

$$T_n = (T_t + T_f) /2 \qquad (B.15)$$

$$T_m = (T_f + T_b) /2 \qquad (B.16)$$

7. Calculate separation factors for stripping and rectification sections, S_{ni} and S_{mi} :

$$S_{ni} = k_i (T_n, P) RV_{Flow} /R_{Flow} \qquad i = 1,2 \qquad (B.17)$$

$$S_{mi} = k_i (T_m, P) RB_{Flow} / R_{Flow} P \qquad i = 1,2 \qquad (B.18)$$

where k_i, i = 1,2   are the vapour and equilibrium constants for stripping and rectification sections in Equations B.17 and B.18.

8. Finally implement reflux flow and reboiler flow setpoints R_{Flow} and RB_{Flow} .

APPENDIX C

%Constrained Multivariable Control Programme for Shell Heavy Oil Fractionator

```
function [y,u]=CRUDE(d1,d2,r1,r2);

%*********************************************************
% Plant Transfer Function (T.F.)
%                   num(s)
% The T.F. is in the form  G(s) = ---------- x exp(-deadtime*s)
%                   den(s)
% Where:
% num(s)  : numerator coeffecients in powers of s
% den(s)  : denominator coefficients in powers of s
% deadtime : dead time
%
% gij=poly2tf(num,den,sampltime,deadtime)
%
% Note: num(s) and den(s) all in descending powers of s
```

% Plant Parameters

```
% tsp   : sampling time [ ZERO FOR CONTINOUS SYSTEMS ]
% tfinal : truncation time for step response models
% tstep  : sampling time for step response models
% nout   : number of outputs [ ONE (1) for SISO SYSTEMS ]
% ny    : number of inputs
%
%
%
%   plant size = |g11  g12 ... g1nu|
%                |g21  g22 ... g2nu|
%                | ... ... ... .. |
%                | ... ... ... .. |
```

```
%                    |gny1  gny2  gnynu|
%
%
%
%
%
%**********************************************************
```

% Plant Definition

```
%g11=poly2tf(12.8,[16.7 1],0,1);
%g21=poly2tf(6.6,[10.9 1],0,7);
%g13=poly2tf(4.8,[8.9 1],0,10);
%g12=poly2tf(-18.9,[21 1],0,3);
%g22=poly2tf(-19.4,[14.4 1],0,3);
%g23=poly2tf(-10.5,[17.2 1],0,4);

g11=poly2tf(4.05,[50 1],0,27);
g21=poly2tf(5.39,[50 1],0,18);
g31=poly2tf(4.38,[33 1],0,20);
g12=poly2tf(1.77,[60 1],0,28);
g22=poly2tf(5.72,[60 1],0,14);
g32=poly2tf(4.42,[44 1],0,22);
g13=poly2tf(5.88,[50 1],0,27);
g23=poly2tf(6.9,[40 1],0,15);
g33=poly2tf(1.14,[27 1],0,0);

tsp=0;
tfinal=500;
tstep=5;
ny=3;


plant=tf2step(tfinal,tstep,ny,g11,g21,g31,g12,g22,g32,g13,g23,g33);
plotstep(plant),pause
```

% Measured Disturbance Definition

```
%gw11=poly2tf(3.8,[14.9 1],0,8);
%gw21=poly2tf(4.9,[13.2 1],0,3);

gw11=poly2tf(1.2,[45 1],0,27);
gw21=poly2tf(1.52,[25 1],0,15);
gw31=poly2tf(1.14,[27 1],0,0);
gw12=poly2tf(1.44,[40 1],0,27);
gw22=poly2tf(1.83,[20 1],0,15);
```

```
gw32=poly2tf(1.26,[32 1],0,0);

wplant=tf2step(tfinal,tstep,ny,gw11,gw21,gw31,gw12,gw22,gw32);
plotstep(wplant)
dplant=wplant;
dmodel=wplant;


model=plant;
```

% No plant/Model Mismatch

```
%gm11=poly2tf(13,[18 1],0,1);
%gm21=poly2tf(6.6,[10.9 1],0,7);
%gm13=poly2tf(6.0,[10 1],0,10);
%gm12=poly2tf(-18.9,[21 1],0,3);
%gm22=poly2tf(-19.4,[14.4 1],0,3);
%gm23=poly2tf(-10.5,[20 1],0,4);


%model=tf2step(tfinal,tstep,ny,gm11,gm21,gm12,gm22,gm13,gm23);

% IF YOU HAVE PLANT/MODEL MISMATCH REMOVE % SIGN FROM
THE PREVIOUS 5
% EQUATIONS AND INSERT % INFRONT OF MODEL=PLANT EQUATION
%**********************************************************
```

% MPC Controller Parameters

```
%  p      : output horizon
%  m      : input horizon   (p>m)
%  yweight : controlled variable weight
%  uweight : manipulated variable weight
%  setpoint: reference trajectory
%  dstep   : disturbance magnitude
%  simtime : simulation time
%  umaxcon : maximum constraint on manipulated variable
%  umincon : minimum constraint on manipulated variable
%  ducon   : constraint on manipulated variable rate-of-change
%          (ABSOLUTE VALUE)
%
%  ulim    : [umincon umaxcon ducon]
%
%**********************************************************

p=15;
m=7;
```

```
ny=3;
nu=3;
setpoint=[r1 r2 0];
dstep=[d1 d2]; % Let this equal zero if you din't have measured disturbance.
ywt=[1 1 1];
uwt=[1 1 1];
simtime=500;
ulim=[-inf -inf -inf inf inf inf 10e6 10e6 10e6]; %Unconstrained case
%ulim=[-0.1 -0.15 inf inf 0.1 100];
ylim=[];
%kmpc=mpccon(model,ywt,uwt,m,p);
[y,u]=cmpc(plant,model,ywt,uwt,m,p,simtime,setpoint,ulim,ylim,[],dplant,dmodel,d
step);
plotall(y,u)
```

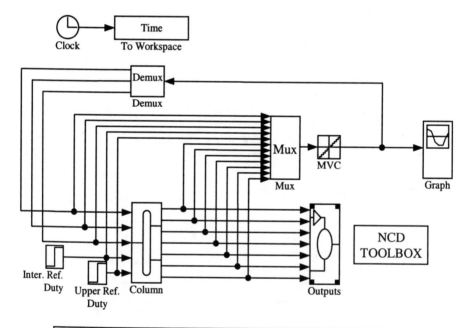

NCD = NON-LINEAR CONTROL DESIGN INITIALIZATION

Double click here to initialize plant data and optimization parameters.
Double click on NCD toolbox block to display constraint windows.

Figure C.1. Non-linear multivariable control SIMULINK® system for Shell heavy oil fractionator

This toolbox provides a graphical user interface for specifying the constraints, tunable variables and plant uncertainty. The simulation system NCD toolbox converts all the constraints, tunable variables and uncertainty into a constrained optimization problem. The optimization performs successive simulation on the SIMULINK® system and changes the tunable variables in an attempt to better meet the time response specifications. The constraints (upper and lower bounds) for this control problem are defined in Section 5.3. The tuning parameters of the control system are given in Table 5.3.

APPENDIX D

D.1 Description and Application of Real-time Optimization (RT-Opt.) Software to Catalytic Reforming Reactor Section

D.1.1 Description

The real time optimization (RT-Opt.) is equation-based closed loop optimization software from Aspentech. It can directly incorporate fundamentally based process models into its system. The process modelling equations derived from the first principal need to be converted into a format suitable to RT-Opt. system. The properly formulated open equation-based models meet this requirement. The fundamental open equation-based models in the RT-Opt. system are fitted to the plant data. On-line scanned measurements such as flows, pressures temperatures and compositions are used to adjust model parameters such as heat transfer coefficients and kinetic constants. The fundamental nature of the models allow physically meaningful parameters to be adjusted to reflect changes in plant equipment performance such as catalyst deactivation or heat exchanger fouling or compressor blade erosion.

The model parameters are adjusted to make the calculated values match the actual measurements prior to each optimization calculation. This is easily accomplished with open equation-based models. Exactly the same models are used in the parameter-fitting step as are used for the optimization step of the RT-Opt. sequence. The continuous scanning capability of the on-line process computer allows the measured data to be analyzed to improve confidence in their accuracy and to ensure that the process is at steady state prior to using the measurements to fit the parameters. The parameterization step is essential to the closed loop optimization process because it ensures that the model matches the current values of the process constraints. The optimization model must know where the plant is relative to the constraints before it can begin to search for the optimum.

D.1.2 Mathematical Algorithm

In RT-Opt., the parameterized open equation-based plant model is manipulated by a mathematical optimization technique such as successive quadratic programming (SQP). The optimization problem should be formatted as:

$$\text{Minimize} \qquad P(\mathbf{X}) \qquad\qquad\qquad\qquad (D.1)$$

$$\text{subject to} \qquad f(\mathbf{X}) = 0 \qquad\qquad\qquad\qquad (D.2)$$

$$\text{and} \qquad \mathbf{X}_{lower} < \mathbf{X} < \mathbf{X}_{upper} \qquad\qquad\qquad (D.3)$$

Here, $\mathbf{X}$ is the vector of model variables and $P(\mathbf{X})$ is the cost-based objective function. It is the function of the model variables, $\mathbf{X}$. The equality constraint, $f(\mathbf{X})$, are the open equation models. The inequality constraints on the model variables, $\mathbf{X}$, are the high and low limits placed on the system to represent plant operating constraints. Mathematical optimization software minimizes the economic cost objective function while driving the model residuals to zero and honoring the process operating constraints.

D.1.3 Application to Catalytic Reforming Reactor Section

The constrained non-linear optimization problem for catalytic reforming reactor section constructed in Section 6.4 was rewritten in the format given in Equations D.1 to D.3.

1. The objective function $P(\mathbf{X})$ needs to be minimized. Equation 6.55 defines the objective function $\mathbf{J}$ to be minimized, subject to:

2. The equality constraint $f(\mathbf{X})$ equal to zero. Equations 6.56 to 6.59 were written in the form $f(\mathbf{X}) = 0$. This is called residual form of equation. The RT-Opt. software manipulates the unknown in these equations such that the residuals are driven to zero.

3. The upper and lower constraints on the weighted average inlet temperature (WAIT)-Equations 6.60 to 6.61 taken into consideration by the optimizer while it minimized the economic cost objective function given by Equation 6.55.

D.2 Implementation Procedure of Real-time Optimization (RT-Opt.)

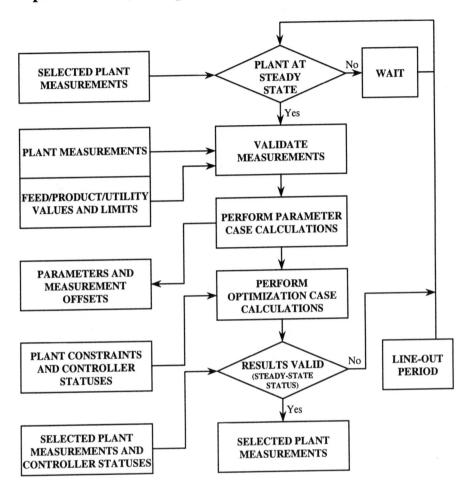

Figure D.1. Implementation procedure of real-time optimization (RT-Opt.) on the catalytic reforming process (reactor section)

APPENDIX E

%Constrained Multivariable Predictive Control for Fluid Catalytic Cracking Process (FCC)

function [y,u]=ansari();

```
%*******************************************************
% Plant Transfer Function (T.F.)
%                      num(s)
% The T.F. is in the form  G(s) = ---------- x exp(-deadtime*s)
%                      den(s)
% Where:
% num(s)   : numerator coeffecients in powers of s
% den(s)   : denominator coefficients in powers of s
% deadtime : dead time
%
% gij=poly2tf(num,den,sampltime,deadtime)
%
% Note: num(s) and den(s) all in descending powers of s
```

% **Plant Parameters**

```
% tsp    : sampling time [ ZERO FOR CONTINOUS SYSTEMS ]
% tfinal : truncation time for step response models
% tstep  : sampling time for step response models
% nout   : number of outputs [ ONE (1) for SISO SYSTEMS ]
% ny     : number of inputs
%
%
%
%   plant size = |g11  g12 ... g1nu|
%                |g21  g22 ... g2nu|
%                | ...  ... ... .. |
%                | ...  ... ... .. |
```

```
%                    |gny1 gny2 gnynu|
%
%
%
%
%
%*********************************************************
```

% Plant Definition

```
g11=poly2tf([0.097*1.7 0.097],[19 6.5 1],0,2);
g21=poly2tf(0,[0 1],0,0);
g31=poly2tf(0,[0 1],0,0);
g41=poly2tf(0,[0 1],0,0);
g51=poly2tf(0,[0 1],0,0);
g12=poly2tf(-0.87,[13 4.9 1],0,2);
g22=poly2tf(0.55,[27 8.7 1],0,4);
g32=poly2tf(0.14,[46 8.5 1],0,11);
g42=poly2tf(0.25,[17 7 1],0,11);
g52=poly2tf(0.66,[2.5 1],0,1);
g13=poly2tf(0.026,[12 1],0,7);
g23=poly2tf(0,[0 1],0,0);
g33=poly2tf(0,[0 1],0,0);
g43=poly2tf(0,[0 1],0,0);
g53=poly2tf(-0.9,[6 1],0,10);
g14=poly2tf([-0.074*4.8 -0.074],[9.3 3.4 1],0,0);
g24=poly2tf([0.74*1.7 0.74],[11 7.3 1],0,2);
g34=poly2tf([0.27*16 0.27],[53 23 1],0,0);
g44=poly2tf(0.7,[3 1],0,0);
g54=poly2tf(1.0,[2 1],0,0);

tsp=0;
tfinal=90;
tstep=3;
ny=5;


plant=tf2step(tfinal,tstep,ny,g11,g21,g31,g41,g51,g12,g22,g32,g42,g52,g13,g23,g3
3,g43,g53,g14,g24,g34,g44,g54);
plotstep(plant),pause

% Measured Disturbance Definition
gw11=poly2tf(3.8,[14.9 1],0,8);
gw21=poly2tf(4.9,[13.2 1],0,3);
wplant=tf2step(tfinal,tstep,ny,gw11,gw21);
```

model=plant;

% No Plant/Model Mismatch

```
%gm11=poly2tf([0.097*1.7 0.097],[19 6.5 1],0,2);
%gm21=poly2tf(0,[0 1],0,0);
%gm31=poly2tf(0,[0 1],0,0);
%gm41=poly2tf(0,[0 1],0,0);
%gm51=poly2tf(0,[0 1],0,0);
%gm12=poly2tf(-0.87,[13 4.9 1],0,2);
%gm22=poly2tf(0.55,[27 8.7 1],0,4);
%gm32=poly2tf(0.14,[46 8.5 1],0,11);
%gm42=poly2tf(0.25,[17 7 1],0,11);
%gm52=poly2tf(0.66,[2.5 1],0,1);
%gm13=poly2tf(0.026,[12 1],0,7);
%gm23=poly2tf(0,[0 1],0,0);
%gm33=poly2tf(0,[0 1],0,0);
%gm43=poly2tf(0,[0 1],0,0);
%gm53=poly2tf(-0.9,[6 1],0,10);
%gm14=poly2tf([-0.074*4.8 -0.074],[9.3 3.4 1],0,0);
%gm24=poly2tf([0.74*1.7 0.74],[11 7.3 1],0,2);
%gm34=poly2tf([0.27*16 0.27],[53 23 1],0,0);
%gm44=poly2tf(0.7,[3 1],0,0);
%gm54=poly2tf(1.0,[2 1],0,0);

%model=tf2step(tfinal,tstep,ny,gm11,gm21,gm31,gm41,gm51,gm12,
gm22,gm32,gm42,gm52,gm13,gm23,gm33,gm43,gm53,gm14,gm24,
gm34,gm44,gm54);
```

% IF YOU HAVE PLANT/MODEL MISMATCH REMOVE % SIGN FROM
THE PREVIOUS MODEL
% EQUATIONS AND INSERT % INFRONT OF MODEL=PLANT EQUATION

%**

% MPC Controller Parameters

% p   : output horizon
% m   : input horizon   (p>m)
% yweight : controlled variable weight
% uweight : manipulated variable weight
% setpoint: reference trajectory
% dstep : disturbance magnitude
% simtime : simulation time
% umaxcon : maximum constraint on manipulated variable

```
%  umincon : minimum constraint on manipulated variable
%  ducon   : constraint on manipulated variable rate-of-change
%            (ABSOLUTE VALUE)
%
%  ulim    : [umincon umaxcon ducon]
%
%******************************************************************

p=15;
m=7;

setpoint=[0 1 5 10 -5];
dstep=0; % Let this equal zero if you don't have measured disturbances
ywt=[1 1 1 1 1];
uwt=[1 1 1 1];
simtime=90;
ulim=[-inf -inf -inf -inf inf inf inf inf 10e6 10e6 10e6 10e6]; %Unconstrained case
%ulim=[-0.1 -0.15 -0.5 -0.6  inf inf inf inf 0.1 100 100 100];
ylim=[];
%kmpc=mpccon(model,ywt,uwt,m,p);
[y,u]=cmpc(plant,model,ywt,uwt,m,p,simtime,setpoint,ulim,ylim);
plotall(y,u,tstep),pause
```

REFERENCES

Agarwal, M. and Seborg, D.E. (1987a). Self tuning controllers for non-linear systems. *Automatica*, 23(2), 209.

Agarwal, M. and Seborg, D.E. (1987b). A multivariable non-linear self-tuning controller. *AIChE Journal*, 33(8), pp.1379.

Ansari, R. M. (1992). On-Line Quality Estimators. *Technical Report*, Shell Refinery, Australia.

Ansari, R.M and Tadé, M.O (1997). Constrained Non-linear Multivariable Control of a Catalytic Reforming Process. *Control Engineering Practice* (CEP), Vol. 6, pp. 695-706.

Ansari, R.M and Tadé, M.O. (1998). Non-linear Model Based Multivariable Control of a Debutanizer. *Journal of Process Control*, Vol.8 (4), pp. 279 -286.

Ansari, R.M., Ghazzawi, A.A. and Bawardi, K. (1997). Model Predictive Multivariable Control on Crude Distillation Unit. *Saudi Aramco Journal of Technology*, Summer Issue, pp:16-21.

Arkun, Y., Hollett. J., Canney, W.M. and Morari, M. (1986). Experimental study of internal model control. *Ind. Eng. Chem. Process Des. Dev.*, 25(1).pp.102-108.

Arulalan,G.R. and Deshpande, P.B. (1986). Synthesizing a multivariable predictive algorithm for noninteractive control. *Proceedings of the American control conference*. Seattle, pp. 1378-1386.

Arulalan,G.R. and Deshpande, P.B. (1987). Simplified model predictive control. *Ind. Eng. Chem. Res.* 26(2), pp. 347-356.

Åström, K. J. (1986). Adaptation, auto-tuning and smart control, Chemical Process Control - CPC III. *Proceedings of the third International Conference on Chemical Process Control, Asilomar, California,* Jan. 12-17, pp. 427-466.

Balchen, J.G., Lie, B. and Solberg, I. (1988). Internal decoupling in non-linear process control. *Modeling, Identification and Control*, 9(3), pp.137-148.

Balchen, J.G., Ljungquist, D. and Strand, S. (1989a). State space model predictive control of a multistage elctrometallurgical process. *Modelling, identification and control*, 10(1), pp.35-51.

Balchen, J.G., Ljungquist, D. and Strand, S. (1989b). Predictive control based upon state space models. *Modelling, identification and control*, 10(2), pp.65-76.

Baratti, R., Bertucco, A., DaRold, A. and Morbidelli, M.(1995). Development of a composition estimator for binary distillation columns: Application to a pilot plant. *Chemical Engineering Science.* Vol.50, nr.10, pp.1541-1550.

Bartusiak, R.D., Georgakis, C. and Reilly, M.J. (1989). Non-linear feedforward/ feedback control structures designed by reference system synthesis. *Chem. Eng. Sci.* 44, pp. 1837-1851.

Bequette, B.W. (1989). A One-Step-Ahead Approach to Non-linear Process Control. *Proc. of the National ISA Meeting*; Philadelphia.

Bequette, B.W. (1990). Process control using non-linear programming techniques. *Lecture notes* in *control and information sciences*, 144, pp57-66.

Bequette, B.W. (1991). Non-linear predictive control of a CSTR using multi-rate sampling disturbance rejection. *Can. J. Chem. Eng.* 69.

Biegler, L.T. and Rawlings, J.R. (1991). Optimization approaches to non-linear model predictive control. *Chemical Process Control – CPCIV*, pp.543-571.

Biegler, L.T. (1990). On solving the fundamental control problem in the presence of uncertainty: A mathematical programming approach. *The second Shell process control workshop, Butterworth.* Stoneham, MA., pp.149-180.

Biegler, L.T. and Rawlings, J.B. (1991). Optimization approaches to non-linear model predictive control. *Chemical Process Control – CPC IV*, pp.543.

Bitmead, R.R., Gevers, M. and Wertz, V. (1990). Adaptive optimal control: The thinking man's GPC, *Prentice-Hall, Englewood Cliffs*, NJ.

Bommannan, D., Srivastava, R.D. and Saraf, D.N. (1989). Modelling of catalytic naphtha reformers. *Canadian J. of Chem. Eng.* 67, pp. 405.

Brosilow, C.B. and Tong, M. (1978). The structure and dynamics of inferential control systems. *AIChE J.* 24(3), pp.492-499.

Brosilow, C.B., Zhao, G.Q. and Rao, K.C. (1984). A linear programming approach to constrained multivariable control. *Proceedings of the third American control conference, San Diego,* June 6-8, 2, pp. 667-674.

Brosilow, C.B. (1979). The structure and design of Smith predictor from the view of inferential control. *Joint automatic control conference proceedings*, Denver, CO.

Brown, M. W., Lee, P. L. and Sullivan, G. R. and Zhou, W. (1990). A Constrained Non-linear Multivariable Control Algorithm. *Trans IChemE*, 68, pp 464-476.

Caldwell, J. M., and Dearwater, J. G. (1991). Model Predictive Control Applied to FCC Units. *Fourth International conference on Chemical Process Control. South Padre Island, Texas,* February 17-22.

Caldwell, J.M. and Martin, G.D. (1987). On-line analyzer predictive control. *Sixth annual control conference*, Rosemont, Illinois, May 19-21.

Callaghan, P.J. and Lee, P.L. (1986). Multivariable predictive control of a grinding circuit. *Proceedings of the third Australian control conference*, Sydney, pp.99-104.

Callaghan, P.J. and Lee, P.L. (1988). An experimental investigation of predictive controller design by principal component analysis. *Chem. Eng. Res. Des.* 66(4), pp.345-356.

Calvet, J.P. and Arkun, Y. (1989). Robust control design for uncertain non-linear systems under feedback linearization. *Proceedings of the conference on decision and control*, Orlando, FL.

Calvet, J.P. and Arkun, Y. (1988a). Feedforward and feedback linearization of non-linear systems with disturbances. *International J. Control*, 48(4), pp.1551-1559.

Calvet, J.P. and Arkun, Y. (1988b). Feedforward and feedback linearization of non-linear systems and its implementation using internal model control. *Ind. Eng. Chem. Res.*, 27(10), pp.1822-1831.

Campo, P. J., Holcomb, T.R., Gelormino, M.S and Morari, M (1990). Decentralized control system design for a heavy oil fractionator – The Shell control problem. *The second Shell process control workshop, Butterworths,* Stoneham, M.A.

Campo, P.J. and Morai, M. (1987). Robust model predictive control. *Proceedings of the American control conference*. Minneapolis, 2, pp.1021-1026.

Chang, T.S., and Seborg, D.E. (1983). A linear programming approach for multivariable feedback control with inequality constraints. *Int. J. Control, 37(3)*, pp. 583-597.

Chen, C.Y. and Joseph, B. (1987). On-line optimization using a two-phase approach: An application study. *Ind. Eng. Chem. Res.*, 26(9), pp.1924-1930.

Chen, C.C. and Shaw, L. (1982). On receding horizon control. *Automatica*, 15, 349.

Cheng, C.M. (1989a). Linear quadratic-model algorithm control method: A controller design method combining the linear quadratic method and the model algorithm control. *Ind. Eng. Chem. Res.* 28(2) pp. 178-186.

Cheng, C.M. (1989b). Linear quadratic-model algorithm control method with manipulated variable and output variable set points and its applications. *Ind. Eng. Chem. Res.* 28(2) pp. 187-192.

Chiu, M.S. and Arkun, Y. (1990). Parameterization of all stabilizing IMC controllers for unstable plants. *Int. J. Control.* 51(2), pp. 329-340.

Clarke, D.W. (1991). Adaptive generalized predictive control. *Proceedings of chemical process control – CPCIV*, pp.395-417.

Clarke, D.W., Mohtadi, C. and Tuffs, P. S (1987a). Generalized Predictive Control: Part I. The basic algorithm. *Automatica, 23(2)*, pp. 137-148.

Clarke, D.W., Mohtadi, C. and Tuffs, P. S (1987b). Generalized Predictive Control: Part II. Extensions and Interpretations. *Automatica, 23(2)*, pp. 149-160.

Cott, B. J., Durham, R. G., Lee, P. L. and Sullivan, G. R. (1989). Process Model Based Engineering. *Computers and Chemical Engineering*, 12, pp 973-984.

Cuthrell, J.E and Biegler, L.T. (1987). On-line optimization of differential-algebraic process systems. *AIChE Journal*, 33(8), pp.1257.

Cuthrell, J.E and Biegler, L.T. (1989). Simultaneous optimization and solution methods for batch reactor control profiles. *Comp. Chem. Eng.*, 13(1/2), pp.49.

Cuthrell, J.E., Rivera, D.E., Schmidt, W.J. and Vegeais, J.A. (1990). Solution to the Shell standard control problem. *The second Shell process control workshop, Butterworth.* Stoneham, MA., pp.27-58.

Cutler, C.R. (1981). Dynamic matrix control of imbalanced systems. *Proceedings of ISA conference and exhibit,* St. Louis, pp.51-56.

Cutler, C.R. (1983). Dynamic matrix control: An optimal multivariable control algorithm with constraints. *Ph.D. thesis,* University of Houston.

Cutler, C.R. and Finlayson, S.G. (1988). Design considerations for a hydrocracker preflash column multivariable constraint controller. *IFAC conference*, Atlanta.

Cutler, C.R. and Hawkins, R.B. (1988). Application of a large predictive multivariable controller to a hydrocracker second stage reactor. *Proceedings of automatic control conference*, Atlanta, GA, pp.284-291.

Cutler, C.R., and Ramaker, B.L. (1980). Dynamic Matrix Control - A Computer Control Algorithm. *Joint Automatic Control Conference Preprints*, Paper WP5-B, San Francisco.

Daoutidis, P. and Kravaris, C. (1989). Synthesis of feedforward/state feedback controllers for non-linear processes. *AIChE Journal*, 35(10), pp.1602.

Dayal, B.S. and MacGregor, J.F.(1997). Recursive exponentially weighted PLS and its applications to adaptive control and prediction. *Journal of Process Control*, Vol.7 (3), pp. 169–179.

De Oliveira, N.M.C and Beigler, L.T. (1995). An extension of Newton-type algorithms for non-linear process control. *Automatica*, 31(2), pp.281.

Denn, M. M. (1986). Process Modelling. *Longam Inc.*, New York.

DMCC. (1993). Dynamic Matrix Optimization and Inferential Calculations. *Technical Report, [DMC] Control Technology*, Houston, Texas.

Douglas, J.M., Jafarey, A and McAvoy, T.J. (1989). Short-cut techniques for distillation column design and control. *Ind. Eng. Chem. Process Des. Dev.*, Vol 18, No. (2), pp 197.

Doyle III, F.J and Hobgood, J.V. (1995). Input-Output linearization using approximate process models. *Journal of Process Control*, Vol.5 (4), pp. 263–275.

Doyle, J. (1982). Analysis of feedback systems with structured uncertainties. *IEEE Proceedings, 129*, AC-26 (1), pp.242-250.

Doyle, J. and Stein, G. (1981). Multivariable feedback design: Concepts for a classical/modern synthesis. *IEEE Trans. Auto. Control*, AC-26 (1), pp.4-16.

Duan. J., Grimble, M.J. and Johnson, M.A. (1997). Multivariabe weighted predictive control. *Journal of Process Control*, Vol.7 (3), pp. 219–235.

Economou, C.G.(1986). An operator theory approach to non-linear controller design. *PhD thesis, California Institute of Technology*.

Economou, C.G., Morari, M., and Palsson, B.O. (1986). Internal Model Control - Extension to Non-linear systems. *Ind. Eng. Chem. Process Des. Dev.*, 25(2), pp. 403-411.

Economou, C.G., Morari, M., Palsson, B.O. (1986). Internal Model Control - Extension to Non-linear systems. *Ind. Eng. Chem. Process Des. Dev.*, 25, pp.403.

Garcia, C.E. (1984). Quadratic/Dynamic Matric Control of non-linear processes. An application to a batch reaction process. *AIChE Annual Meeting*, paper no.82f, San Francisco, C. A.

Garcia, C.E., Morari, M. (1982). Internal Model Control. 1. A Unifying Review and some new Results. *Ind. Eng. Chem. Proc. Des. Dev., 21(2)*, pp. 308-323.

Garcia, C.E., Morari, M. (1985a). Internal Model Control. 2. Design Procedure for Multivariable Systems. *Ind. Eng. Chem. Proc. Des. Dev., 24(2)*, pp. 472-484.

Garcia, C.E., Morari, M. (1985b). Internal Model Control. 3. Multivariable Control Law Computation and Tuning Guidelines. *Ind. Eng. Chem. Proc. Des. Dev., 24(2)*, pp. 484-494.

Garcia, C.E., Prett, D.M., and Morari, M. (1989). Model Predictive Control: Theory and Practice - a survey. *Automatica, 25*, pp. 335-348.

Garcia, C.E. and Morshedi, A.M. (1986). Quadratic programming solution of dynamic matrix control (QDMC). *Chem. Engng. Commun.* 46, 73-87.

Garcia, C.E., Prett, D.M. and Morari,M. (1989). Model predictive control: Theory and practice- a survey. *Automatica.* 25(3), pp.335-348.

Gattu, G. and Zafiriou, E. (1992). Non-linear quadratic dynamic matrix control with state estimation. *Ind. Eng. Chem.*, 31(4), pp.1096-1104.

Gill, P.E., Murray, W., Saunders, M.A., and Wright, M. H. (1986). User's Guide for NPSOL (Version 4.0): A Fortran Package for Non-linear Programming. *Technical Report SOL 86-2, Stanford University,* Stanford, California.

Grimble, M.J., De La Salle, S. and Ho, D. (1989). Relationship between internal model control and LQG controller structure. *Automatica*, 25(1),pp.41-53.

Grimm, W.M., Lee, P.L. and Callaghan, P.J. (1989). Practical robust predictive control of a heat exchanger network. *Chem. Eng. Commun.* 81, pp.25-53.

Grosdidier, P., Mason, A., Aitolahti, A., Heinonen, P., and Vanhamaki, V (1993). FCC Unit Reactor-Regenerator Control. *Computers Chem. Engng.* 17(2), pp. 165-179.

Häggblom, K.E. (1996). Combined internal model and inferential control of a distillation column via closed-loop identification. *Journal of Process Control*, Vol.6 (4), pp. 223-232.

Harris, T.J and MacGregor, J.F. (1987). Design of multivariable linear quadratic controllers using transfer functions. *AIChE Journal.* 33(9),pp.1481-1495.

Henson, A.M., Seborg, D.E. (1989). Extension of Non-linear Coupling Methods to include Feedback Linearization. Presented at the *AIChE Annual Meeting*, San Francisco, C. A.

Henson, M.A. and Seborg, D.E. (1989). A unified differential geometric approach to non-linear process control. *AIChE Annual meeting*, San Francisco.

Henson, M.A. and Seborg, D.E. (1990). Non-linear control strategies for continuous fermentors. *Proceeding of the American control conference*, San Diego, CA. pp.2723-2728

Henson, M.A. and Seborg, D.E. (1993). Theoretical analysis of unconstrained non-linear model predictive control. *Int. J. Control*, 58, pp.1053.

Hernández, E. and Arkun, Y. (1993). Control of non-linear systems using polynomial ARMA models. *AIChE Journal* , 39(3), pp.446-460.

Hmood, S.A., and Prasad, R.M. (1987). Generalized approach for model algorithm control. *Int. J. Systems Sci.* 18, pp.1395-1410.

Holt, B.R. and Morari, M. (1985a). Design of resilient processing plants – IV. The effect of right half plane zeroes on dynamic resilience. *Chem. Eng. Sci.* 40(1), pp.59-74.

Holt, B.R. and Morari, M. (1985b). Design of resilient processing plants – V. The effect of dead time on dynamic resilience. *Chem. Eng. Sci.* 40(7), pp.1229-1237.

Hsie, W.L. (1989). Modelling, Simulation and Control of a Crude Tower, *Ph.D. dissertation*, University of Maryland.

Huang, H., and Stephanopoulos, G (1985). Adaptive Design for model-based controllers. *Proceedings of the American Control Conference,* Vol.3, pp. 1520-1527.

Huq, I., Morari, M. and Sorensen, R.C. (1995). Modification to model IV fluid catalytic cracking units to improve dynamic performance. *AIChE Journal., 41*, pp. 1481.

InfoPlus Product Family. (1992). A Real-Time Data Management and Control System for Hydrocarbon Processing Plant. *SetPoint Inc.*, Houston, Texas.

Ishida M. and Zhan, J. (1995). Neural model predictive control (NMPC) of a distributed parameter crystal growth process. *AIChE. J.*, 41, pp. 2333-2336.

Jafarey, A. and McAvoy, T. (1988). Steady-State non-interacting controls for distillation columns: Analytical study. *Ind. Eng. Chem. Process Des. Dev.*, 19, pp.114-117.

Jang, S., Joseph, B. and Mukai, H. (1987). Control of constrained multivariable non-linear process using a two-phase approach. *Ind. Eng. Chem. Res.*, 26, pp.809.

Joseph, B. and Brosilow, C. (1978). Steady state analysis and design. *AIChE J.* 24(3), pp.485-492.

Jun, K.S., Rivera, D.E., Elisante, E. and Sater, V.E. (1996). A computer-aided design tool for robustness analysis and control-relevant identification of Horizon Predictive Control with application to a binary distillation column. *Journal of Process Control*, Vol.6 (2/3), pp. 177 -186.

Kantor, J.C. (1989). Non-linear sliding-mode controller and objective function for surge tanks. *Int. J. Control*, 50(5), 2025.

Keeler, J.D., Martin, G., Boe, G., Piche, S., Mathur U., and Johnson D. (1996). The Process Perfecter: The next step beyond Multivariable Control and Optimization. *Pavilion Technologies Inc. Technical Report, Austin, Texas.*

Khambanonda, T.A., Palazoglu, A., Romagnoli, J.A. (1990). A study of the controller tuning for stabilizing non-linear feedback systems based on generalized models. *Proceedings of the American control conference*, San Diego, CA., pp.2752-2757.

Kozub, D.J., MacGregor, J.F. and Harris, T.J. (1989). Optimal IMC inverses: Design and robustness considerations. *Chem. Eng. Sci.* 44(10), pp.2121-2136.

Kozub, D.J., MacGregor, J.F. and Wright, J.D. (1987). Application of LQ and IMC controllers to a packed bed reactor. *AIChE Journal.* 33(9), pp.1496-1506.

Kravaris, C. (1988). Input /output linearization: A non-linear analog of placing poles at process zeros. *AIChE Journal*, 43(11), pp.1803-1812.

Kravaris, C. and Daoutidis, P. (1990). Non-linear state feedback control of second order nonminimum-phase non-linear systems. *Computers and Chemical Engineering*, 14(4/5), pp.439-449.

Kravaris, C. and Wright, R.A. (1989). Deadtime compensation for non-linear processes. *AIChE Journal*, 35(9), pp.1535-1543.

Kresta, J., Marlin, T.E. and MacGregor, J.F. (1990). Choosing inferential variables using projection to latent structure (PLS) with application to multicomponent distillation. Paper 23F, *AIChE Annual meeting*, Chicago.

Kurihara, H. (1967). Optimal Control of Fluid Catalytic Cracking Processes. *ScD. Thesis, MIT.*

Lee, J.W., Ko, Y.C., Jung, Y.K. and Lee, K.S (1997). A Modelling and Simulation Study on a Naphtha Reforming Unit with a Catalyst Circulation and Regeneration System. *Computers and Chemical Engineering,* vol 21, suppl., pp. S1105-S1110.

Lee, J.H., Gelormino, M.S. and Morari, M. (1991). Model predictive control of multi-rate sampled data systems: A state-space approach. *Journal of Process Control.*

Lee, J.H., Gelormino, M.S., Lundstrom, P. and Morari, M. (1990). Model predictive control of multi-rate sampled data systems. paper 4D, *AIChE Annual meeting*, Chicago.

Lee, J.H., Morari, M. and Garcia, C.E. (1994) State-space interpretation of model predictive control. *Automatica*, 30, pp.707-717.

Lee, P. L. (1991). Direct Use of Non-linear Models for Process Control. *Chemical Process Control - CPCIV*, pp 973-984.

Lee, P.L., and Sullivan, G.R. (1988). Generic Model Control. Computers and Chemical Engineering., 12(6), pp.573-580.

Lee, P.L., and Sullivan, G.R. and Zhou, W. (1989). Process/Model Mismatch Compensation for Model-based Controllers. *Chem. Eng. Comm.*, 80, pp. 33-51.

Lee, P.L., and Sullivan, G.R. and Zhou, W. (1990). A New Multivariable Deadtime Control Algorithm. *Chem. Eng. Comm.*, 91, pp. 49-63.

Levien, K.L. and Morari, M. (1987). Internal model control of coupled distillation columns. *AIChE Journal.* 33(1), pp.83-98.

Li, W.C., and Biegler, L.T. (1988). Process Control Strategies for Constrained Non-linear Systems. *Ind. Eng. Chem. Res.*, 27(8), pp. 1421-1433.

Li, W.C. and Biegler, L.T.(1989). Multistep Newton-type control strategies for constrained non-linear process. *Chem. Eng. Res.*, 67, pp.562.

Little, D.L., and Edgar, T.F. (1986). Predictive Control using Constrained Optimal Control. *Proceedings of the American Control Conference,* Seattle, June 18-20, 3, pp. 1365-1371.

Liu, S.L. (1967). Noninteracting Process Control. *Ind. Eng. Chem. Process Des. Dev.*, 6, pp. 460-468.

Lu, Z. and Holt, B.R. (1990). Non-linear robust control: Table look-up controller design. *Proceedings of the American control conference,* San Diego, CA. pp. 2758-2763.

Lundstrom, P., Lee, J.H., Morari, M and Skogestad, S. (1995). Limitations of dynamic matrix control. *Computers chem. Engng.*, Vol.19, no.4, pp.409-421.

Luyben, W.L.(1968). Non-linear feedforward control of chemical reactors. *AIChE Journal,* 14(1), 37.

Magalhaes, M.V., and Odloak, D. (1995). Multivariable quality control of a crude oil fractionator. *IFAC Dynamics and control of chemical reactors (DYCORD '95), Copenhagen,* Denmark, pp. 469-474.

Malik, S. A . (1991). Identification of Mathematical Models for Inferential Control in Refineries, *Technical Report, Amoco Oil Company*; Chicago.

Manousiouthakis, V. (1990). A game theoretic approach to robust controller synthesis. *Com. Chem. Eng.*, 14(4/5), pp.381-389.

Marchetti, J. L., Mellichamp, D.A and Seborg, D.E (1983). Predictive Control Based on Discrete Convolution Models. *Ind. Eng. Chem. Proc. Des. Dev.*, 22, pp. 488-495.

Mark, M., John, E.C. and Michael, C.D. (1987). Advanced Control of a Catalytic Reforming Unit. *NPRA Annual Meeting, Convention Center, San Antonio, Texas,* March 29-31.

Martin, G.D., Caldwell, J.M. and Ayral, T.E. (1986). Predictive control applications for the petroleum refining industry. *Japan petroleum institute – petroleum refining conference,* Tokyo, Japan, October 27-28.

Martin, G.D., Mahoney, J. D., and Kliesch, H.C (1985). Rigorous Simulation used to determine FCC Computer Control Strategies. *Sixth Annual Fluid Catalytic Symposium,* Munich, Germany, May 21-23.

Maurath, P.R. Mellichamp, D.A. and Seborg, D.E. (1985a). Predictive controller design for SISO systems. *Proceedings of the 1985* American *control conference*, 3, pp. 1546-1552.

Maurath, P.R. Seborg, D.E. and Mellichamp, D.A. (1985b). Predictive controller design by principle component analysis. *Proceedings of the 1985* American *control conference*, 3, pp. 1059-1065.

Maurath, P.R. Seborg, D.E. and Mellichamp, D.A. (1988). Predictive controller design by principle component analysis. *Ind. Eng. Chem. Res.* 27(7), pp. 1204-1212.

Mayne, D.Q. and Michalska, H. (1990). Receding horizon control of non-linear systems. *IEEE Trans. Automatic Control.* AC-35, pp.814.

McAvoy, T., Arkun, Y. and Zafiriou, E. (1989). Model-based process control: *Proceedings of the IFAC workshop.* Pergamon Press, Oxford.

McDonald, K. (1987). Performance Comparison of Methods for High Purity Distillation Control. *AIChE Spring National Meeting, Houston, Texas.*

McDonald, K.A. and McAvoy, T.J and Tits, A. (1986). Optimal averaging level control. AIChE J.,32(1), 75.

McFarlane, R.C., Reineman, R.C., Bartee, J. F., and Georgakis, C (1993). Dynamic Simulator for a model IV fluid catalytic cracking unit. *Computers Chem. Engng.* 17, pp. 275-285.

Meadows, E., Henson, M.A., Eaton, J.W., and Rawlings, J.B. (1995). Receding horizon control and discontinuous feedback stabilization. *Int. J. Control,* 62(5), pp.1217.

Meadows, E.S. and Rawlings, J.B. (1990). Simultaneous model identification and control of a semi-batch chemical reactor. *Proceedings of the American control conference,* San Diego, CA, pp.1656-1659.

Mehra, R.K., and Rouhani, R. (1980). Theoretical considerations on model algorithmic control for nonminimum phase systems. *Joint automatic control conference.* Paper TA8-B, San Francisco.

Mehra, R.K., and Rouhani, R., Eterno, J., Richalet, J. and Rault, A. (1981). Model algorithmic control: Review and recent development. Chemical and process control. *Proceedings of the engineering foundation conference,* pp.287-309.

Mehra, R.K., Rouhani, R. and Praly, L. (1980). New theoretical developments in multivariable predictive algorithm control. *Joint automatic control conference.* Paper FA9-B, San Francisco.

Mehra, R.K., Rouhani, R., Eterno, J., Richalet, J. and Rault, A. (1982). Model algorithm control: Review and recent development.. *Engineering foundation conference on chemical process control II.* pp. 287-310.

Mehra, R.K., Rouhani, R., Rault, A., Reid, J.G. (1979). Model algorithm control: Theoretical results on robustness. *Joint automatic control conference.* Denver, pp. 387-392.

Mejdell, T. and Skogetad, S. (1990). Composition control of distillation columns using multiple temperature measurements. paper 23G, *AIChE Annual Meeting,* Chicago.

Mills, P.M., Zomaya, A.Y. and Tadé, M.O. (1995). *Neuro-Adaptive Process Control: A Practical Approach. John Wiley and Sons Ltd., England.*

Miller, S. M. and Rawlings, J. B. (1994). *AIChE J.,* 40(8), pp. 1312.

Moore, C., Hackney, J. and Canter, D. (1986). Selecting sensor location and type of multivariate processes. *Shell Process Control Workshop,* Butterworths, Toronto.

Morari, M. (1987). Robust Process Control. *Chem. Eng. Res. Des.* 65, pp.462-479.

Morari, M. and Doyle, C.J.(1986). A unifying framework for control system design under uncertainty and its implications for chemical process control. *Proceedings of the conference on Chem. Process Control III,* Asilomar, CA.

Morari, M. and Zafiriou, E. (1989). Robust Process Control. *Prentice Hall, Englewood Cliffs,* NJ.

Morari, M., and Lee, J.H (1991). Model Predictive Control: The Good, the Bad, and the Ugly. *Chemical Process Control Conf.- CPCIV,* South Padre Island, Texas, Feb. 17-22, pp. 419-444.

Moro, L.F., and Odloak, D (1995). Constrained Multivariable Control of Fluid Catalytic Cracking Converters. *Journal of Process Control,* 5(1), pp. 29-39.

Morshedi, A.M., Cutler, C.R and Skrovanek, T.A. (1985). Optimal solution of Dynamic Matrix Control with linear programming techniques (LDMC). *Proceedings of American control conference,* Boston, pp.199-208.

Morshedi, A.M. (1986). Universal dynamic matrix control. *Chemical Process Control CPC-III.* Asilomar, CA. pp.547-577.

Morshedi, A.M., Lin, H.Y. and Luecke, R.H. (1986). Rapid computation of the Jacobian matrix for optimization of non-linear dynamic processes. *Comp. Chem. Engng.,* 10(4),pp.367-376.

Murtagh, B.A. and Saunders, M.A. (1987). MINOS 5.1 User's Guide. *Technical Report SOL83-20R*, Stanford University.

Muske, K., Young, J., Grosdidier, P and Tani, S. (1991). Crude unit product quality control. *Computers and Chemical Engineering*, Vol.15, No.9, pp.629-638.

Newell, R.B. and Lee, P.L. (1989). Applied process control, *Prentice Hall, Englewood Cliffs*, New Jersey.

Nikolaou,M. and Manousiouthakis, V. (1989). A hybrid approach to non-linear system stability and performance. *AIChE Journal*, 35(4), pp.559.

Nikravesh M., Farell, A.E., Lee, C.T. and VanZee, J.W. (1995). Dynamic matrix control of diaphragm-type chlorine/caustic electrolysers. *Journal of Process Control*, Vol.5 (3), pp. 131–136.

Noh, S.B., Kim, Y.H., Lee, Y.I. and Kwon, W.H. (1996). Robust generalized predictive control with terminal output weightings. *Journal of Process Control*, Vol.6 (2/3), pp. 137 -144.

O'Connor, D.L., Grimstad, K. and McKay, J. (1991). Application of a single multivariable controller to two hydrocracker distillation columns in series. *ISA meeting*, Anaheim.

Ogunnaike, B.A. and Adewale, K.E.P. (1986). Dynamic matrix control for process systems with time varying parameters. *Chem. Eng. Commun.* 47, pp.295-314.

Ohshima, M., Ohno, H., Hashimoto, I., Sasajima, M., Maejima,M.,Tsuto,K. and Ogawa,T. (1995). Model predictive control with adaptive disturbance prediction and its application to fatty acid distillation column control. *Journal of Process Control*, Vol.5 (1), pp. 41–48.

Palazoglu, A., Fruzzetti, K.P., Romagnoli, J.A. and McDonald, K.A. (1989). Robust multivariable predictive control: A linear programming approach. *Annual AIChE meeting*. Paper no. 21f, San Francisco, CA.

Papadoulis, A.V. and Svoronos, S.A. (1987). Adaptive MDC control. *Proceedings of American control conference*. Minneapolis, pp.2038-2044.

Parrish, J. R. (1985). Non-linear Inferential Control. *Ph.D thesis*, Case Western Reserve University, Ohio.

Parrish, J.R. and Brosilow, C.B. (1986). Non-linear inferential control. *AIChE Journal*. 34(4), pp.633-644.

Patwardhan, A.A., Rawlings, J.R. and Edgar, T.F. (1988). Non-linear Predictive Control Using Solution and Optimization. Presented at the *AIChE National Meeting*, Washington D.C.

Patwardhan, A.A., Rawlings, J.R. and Edgar, T.F. (1990). Non-linear Predictive Control. *Chem. Eng. Comm.*, 87,pp.123.

Patwardhan, A.A. and Edgar, T.F. (1993). Non-linear Model Predictive Control of a Packed Distillation Column. *Ind. Eng. Chem. Res.*, 32 (10), pp.2345-2356.

Patwardhan, A.A., Wright, G.T. and Edgar, T.F. (1992). Non-linear Model Predictive Control of Distributed-Parameter Systems. *Chem. Eng. Sc.*, 47 (4), pp.721-735.

Perry, R.H. and Green, D. (Eds) (1984). *Chemical Engineer's Handbook*, Sixth Edition, 4-68. McGraw Hill, New York.
Pollard, J.F and Brosilow, C.B (1985). Model Selection for Model Based Controller. *Proceedings of the 1985 American control conference*, 3, pp. 1286-1292.

Prett, D.M. and Garcia, C.E. (1987). Design of Robust Process Controllers. *Proceedings of IFAC 10th Triennnial World Congress, Munich*, 2, pp. 275-280.

Prett, D.M. and Morari, M. (1987). The Shell Process Control Workshop. *Butterworths*, Stoneham, M.A.

Prett, D.W., and Gillette, R.D (1979). Optimization and Constrained Multivariable Control of a Catalytic Cracking Unit. *AIChE National Meeting*, Houston, Texas.

Prett, D.M. and Gillette, R.D. (1980) *Proceedings of joint automatic control conference*. San Francisco, CA.

Prett, D.M., and Garcia C.E. (1988). Fundamental Process Control. *Butterworths*, Boston,M.A.

Rawlings, J.B. and Muske, K. (1993). The stability of constrained receding horizon control. *IEEE Trans. Automatic Control*. AC-38, pp.1512.

Rawlings, J.B., Meadows, E.S. and Muske, K.R (1994). Non-linear Model Predictive control. A tutorial and survey. *ADCHEM '94*. Kyoto, Japan.

Rawlings, J.B., Jerome, N.F., Hamer, J.W. and Bruemmer, T.M. (1989). End-point control in semi-batch chemical reactors. *Proceedings of the IFAC symposium on dynamics and control of chemical reactors, distillation columns and batch processes*, DYCORD+ '89, Maastricht, The Netherlands.

Ray, W.H. (1982). New approaches to the dynamics of non-linear systems with implications for process and control system design. *Chemical process control II*, pp.246-267.

Ray, W.H. (1983). Multivariable process control - A survey. *Comp. Chem. Engng.*, 7(4),pp.367

Read, N.K. and Ray, W.H. (1998). Application of non-linear dynamic analysis in the identification and control of non-linear systems – Simple dynamics. *Journal of Process Control*, Vol.8 (1), pp. 1-15.

Richalet, J., Rault, A., Testud, J.L. and Papon, J. (1978). Model predictive heuristic control. Application to industrial processes. *Automatica,* 14, pp. 413-428.

Ricker, N.L. (1985). Use of Quadratic Programming for Constrained Internal Model Control. *Ind. Eng. Chem. Process Des. Dev.*, 24(4), pp. 925-936.

Ricker, N.L. (1990). Model predictive control with state estimation. *Ind. Eng. Chem. Res.*, 29, pp.374-382.

Riggs, J.B. (1990). Non-linear Process Model Based Control of a Propylene Sidestream Draw Column. *Ind. Eng. Chem. Res.*, Vol 29, No. 11.

Riggs, J.B., Rhinehart, R. R. (1988). Comparison Between Process Model Based Controllers. *Proceedings of the American Control Conference*, Atlanta, G. A.

Rivera, D.E., Morari, M., and Skogetad, S. (1986). Internal model control: PID controller design. Ind. Eng. Chem. Process Des. Dev. 25(1), pp.252-265.

Rotea, M.A. and Marchetti, J.L. (1987). Internal model control using the linear quadratic regulatory theory. *Ind. Eng. Chem. Res.* 26(3), pp. 577-581.

Rouhani, R. and Mehra, R.K (1982). Model Algorithmic Control (MAC): Basic Theoretical Properties. *Automatica,* 18(4), pp. 401-414.

Rovaglio, M., Ranzi, E., Biardi, G., Fontana, M. and Domenichini, R. (1990). Rigorous dynamics and feedforward control design for distillation processes. *AIChE Journal*, 36(4), pp. 576-586.

Safonov, M.G. (1980). Stability and robustness of multivariable feedback systems. *The MIT press*, Cambridge, MA.

Seborg, D.E., Edgar, T.F; Shah, S.L (1986). Adaptive Control Strategies for Process Control: A Survey. *AIChE Journal,* 32(6), pp. 881-913.

Seborg, D.E. (1987). The prospects for advanced process control. *Proceedings of IFAC 10th triennial world congress*, Munich, 2, pp.281-289.

Shinnar, R. (1986). Impact of model uncertainties and non-linearities on modern controller design: Present status and future goals. *Chemical Process Control-CPCIII*, pp.53-93.

Shinskey, F.G. (1988). Process Control Systems. Application, Design and Adjustment. *Third Edition*, McGraw-Hill Inc.

Sistu, P.B., Gopinath, R.S. and Bequette, B.W. (1993). Computational issues in non-linear predictive control. *Comp. Chem. Engng.*, 17(4),pp.361-366.

Sistu, P.B. and Bequette, B.W. (1990). Process identification using non-linear programming techniques.. *Proceedings of the American control conference,* pp.1534-1539.

Skogestad, S. and Mejdell, T. (1990). Output estimation for ill-conditioned plants using multiple secondary measurements: High purity distillation. *Annual AIChE Conference Proceedings*, Chicago.

Signal, P.D., and Lee, P.L (1992). Generic Model Adaptive Control. *Chem. Eng. Comm,* Vol. 115, pp. 35-52.

Smith, O.J. M. (1957). Closer control of loops with deadtime. *Chemical Engineering Progress*, 53(5), pp.217-219.

Smith, B.D., Brinkley, W.K.(1960). General Shortcut Equation for Equilibrium Stage Processes. *AIChE J.* Vol. 6, No. (3), pp 446.

Smith, J.M. (1970). *Chemical Engineering Kinetics*, 2nd ed; McGraw Hill, New York.

Spangler, M.V. (1994). Uses of Process Insights at a Refractory Gold Plant. *Pavilion Users Conference,* October 17, Texas.

Sriniwas, G.R. and Arkun, Y. (1997). Control of the Tennessee Eastman process using input-output models. *Journal of Process Control*, Vol.7 (5), pp. 387–400.

Stephanopoulos, G. (1984). Chemical Process Control. An introduction to theory and practice. *Prentice-Hall, Englewood Cliffs,* New Jersey.

Suarez-Cortez, R. Alvarez-Gallegos, J. and Gonzalez-Mora (1989). Sliding controller design for a non-linear fermentation system. *Biotechnology and bioengineering*, 33, pp.377-385.

To, L.C., Tadé, M.O., Kraetzl, M., and LePage, G.P. (1995). Non-linear control of a simulated industrial evaporation process. *Journal of Process Control*, Vol.5 (3), pp. 173–182.

Ungar, L.H., Hartman, E.J., Keeler, J.D. and Martin, G.D. (1995). Process modelling and control usingneural networks. *Proceedings of the intelligent systems in process engineering conference (ISPE)*, Snowmas, pp. 1–11.

Wong, S.K.P. (1985). Control and modelling strategies for non-linear systems with time delays. *Ph.D. thesis, University of California*, Santa Barbara.

Wong, S.K.P. and Seborg, D.E. (1986a). Low-order, non-linear dynamic models for distillation columns. *Proceedings of the American control conference.*

Wong, S.K.P. and Seborg, D.E. (1986b). Control strategies for non-linear multivariable systems with time delays. *Proceedings of the American control conference*, pp.1023.

Wright, J.D., MacGregor, J.F., Jutan, A., Tremblay, J.P. and Wong, A. (1977). Inferential control of an exothermic packed bed reactor. *Proc. Joint Aut. Cont. Conf.*, San Francisco, pp.1516-1522.

Zafiriou, E. and Morari, M. (1986). Design of the IMC filter by using the structured singular value approach. *Proceedings of the American control conference*, Seattle. pp.1-6.

Zafiriou, E. and Morari, M. (1987). Robust H_2-type IMC controller design via the structured singular value. *Proceedings of the IFAC 10th triennial world congress.* Munich. pp.259-264.

Zafiriou, E. and Morari, M. (1988). A general controller synthesis methodology based on the IMC structure and the H_2 -, H_∞- and μ-optimal control theories. *Computers and Chem. Engng.* 12(7), pp.757-765.

Zheng, Y.Y. (1994). Dynamic Modelling and Simulation of a Catalytic Cracking Unit. *Computers Chem. Engng.* 18(1), pp. 39-44.

Zhou, W. and Lee, P.L. (1995). Robust non-linear process control: The necessity of integral control. *Proceedings of Control '95 conference*. Melbourne, pp.61-65.

INDEX